Hitting the **BRAKES**

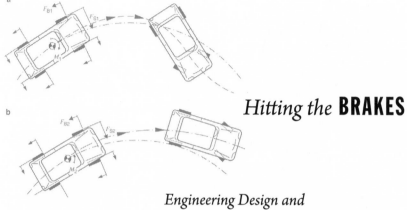

Hitting the BRAKES

Engineering Design and
the Production of Knowledge

Ann Johnson

Duke University Press
Durham and London
2009

© 2009 Duke University Press

All rights reserved

Printed in the United States of America on
acid-free paper ∞

Designed by Heather Hensley

Typeset in Arno Pro by Keystone Typesetting

Library of Congress Cataloging-in-Publication
data appear on the last printed page of this
book.

Contents

Preface

— — —

This book begins with an introduction to the practices of engineering design. Compared to scientific practices, the work of engineers is far less frequently studied and theorized. When concepts from the history and philosophy of science are applied to the history and philosophy of engineering, strange distortions occur. The introduction attempts to clarify such distortions and to legitimate engineering knowledge by its own standards. Doing so requires a recognition that engineering knowledge is a genuine kind of knowledge, not merely the application of already existent scientific knowledge. Since knowledge is seen as something engineering communities produce, this chapter also explores the entanglement between engineering's social structures and its knowledge. This chapter introduces the concept of a knowledge community, which is a small, informal community of practitioners that aggregates around a simultaneously developing question. As this question evolves, so does the knowledge community's informal membership—practitioners drop in and out. In this book knowledge and community are approached symmetrically, with neither being privileged as the precursor of the other. The historical case study of antilock braking systems (ABS) makes this conceit possible, since the social and epistemological dimensions of ABS design need not be separated. The introduction also investigates the messy relationship between technological artifacts and ideas, looking at how ideas beget objects and how objects bear ideas.

So as to orient the reader to the history of ABS and its community, chapter 2 provides a narrative of the development of ABS between World War II and the introduction of integrated ABS systems in the late 1970s. It is important to see the full arc of the story in order to understand the nonlinear ways that the knowledge community and its artifacts developed. The development of ABS is a complicated story involving many different actors, institutions, and artifacts embedded in an array of dynamic national, economic, and cultural contexts over more than twenty-five years. In this overview narrative in chapter 2, I map the relatively circuitous route of ABS development; in doing so I also lay out the social, economic, and cultural background of ABS. This chapter provides something of the standard story of engineering work and product development against which I contrast the much more complex processes of knowledge production and community formation. In subsequent chapters my focus is thematic or institutional rather than narrative and chronological.

Chapter 3 returns to the earliest post–World War II investigations into the skidding problem and focuses on the role of government institutions and the interplay between agencies and the private sector. Between 1949 and 1952 several aircraft brake producers introduced antiskid devices for airplanes. This coincided with statistical investigations into the frequency and injury rate of automobile accidents in Great Britain. By the early 1950s automobile safety was a growing concern in the developed world. The seemingly endless proliferation of cars on the road and the number of miles driven did not bode well for reducing accidents without some kind of intervention from both car companies and governments. Throughout the 1950s skidding seemed to be causing not only more, but more serious accidents as well, as reported in British police reports. The British Road Research Laboratory (RRL) defined the problem of skidding as the "interaction between the road, tyre, vehicle and driver" and brought together specialists within the RRL from each of these areas to try to find solutions to skidding. They discovered quickly that specialists from each different area understood skidding in different ways, so solving the problem depended on developing a common definition of the problem of skidding. Defining skidding required a multidisciplinary community devoted to finding a common language— this was the genesis of the knowledge community which, led by the RRL, gathered for the first time at the International Skid Prevention Conference in 1958 at the University of Virginia. This conference brought together specialists in all of the areas connected to skidding, from tire design to driver psychology, and provided an environment in which problem-oriented subcommittees could share information and collectively define the problem of skidding. The most important research at the RRL led to the development of the first antiskid device for automobiles and pioneered a partnership between the British government and the Dunlop tire company. The work of the RRL became a model for other agencies interested in working on "interaction problems," which were often found in the newly investigated areas of safety improvement. The chapter ends with a case study of the way patents were used to facilitate the sharing of knowledge produced at the RRL.

Chapter 4 examines the formation and development of a metrology and research technology knowledge community that formed around the questions of how and what to measure in order to determine and define skidding. Engineers involved in the design of braking systems, particularly those in proportioning and limiting valve design, and inventors of scientific instruments such as dynamometers worked with engineers measuring the interaction between the road and the vehicle relative to skidding. Old systems and protocols were modified and new systems invented, and these machines and methods were diffused throughout related communities of engineers. The definition of "acceptable" performance also emerged from the community of research technologists, since what could be measured defined performance categories. This chapter includes an inquiry into the problem of measuring qualitative concepts, such as "handling" and "feel." While metrology often seemed at first glance a quantitative subject, testing automotive components involved wrestling with the interaction problem, how to measure how drivers actually used systems, which often did not correlate directly with bench measurements of performance. This chapter concludes with a discussion of computer simulations of precision measurement, an area that united instrumentation with the next chapter's subject, vehicle dynamics.

Measuring a vehicle's braking performance would be a meaningless task without mathematical models to use the empirical data produced by those measurements. Chapter 5 examines the development of mathematical models and algorithms of vehicle dynamics, a field of great interest in the late 1950s through the mid-1960s. Engineers and mathematicians from both automotive and aircraft design were interested in constructing theories of how and why vehicles behaved the way they did. The hope was that a mathematical understanding of vehicle dynamics would lead to the construction of mathematical models of the handling characteristics of vehicles. But research in vehicle dynamics never fulfilled its promise. Theoretical modeling of skidding looked like the most fruitful path to understanding skidding and its complexity in the late 1950s, but by the mid-1960s these models appeared hopelessly complicated for real-time analysis. This failure shifted the focus of the participants in the skid prevention community back to the development of instrumentation and research technology, which by its nature oper-ated in real time. Ironically, the interest in mathematical models of vehicle dynamics reemerged in the mid-1970s with the development of electronic adaptive control systems. Algorithms were a necessary com-ponent of electronic control systems, and microprocessors changed the scale of real-time analysis, requiring sophisticated mathematical models of vehicle dynamics.

Chapters 6 and 7 examine the market introduction of a series of antilock systems beginning in 1966. Chapter 6 focuses on American sys-tems and their failures. Early systems were plagued by high cost, limited production, and serious performance problems. By this time, system designers agreed that antilock systems had two desiderata: shortening braking distances; and allowing drivers to retain directional control (i.e., steering). Given that cutting edge automotive technologies are usually introduced at the high end of the market, where consumers expect optimal performance, the performance limitations of the early systems were disastrous. Early systems were not reliable and often did not im-prove performance. These initial technologies were flawed because of weaknesses and mistakes in the design criteria and parameters that the systems were designed to meet. Rushed into production, none of these systems, produced by Kelsey-Hayes and Ford and Bendix and Chrysler, were able to reduce braking distance; the Ford and Kelsey-Hayes system did not even allow the driver to retain steering control. After the market failure of these systems, engineers had to reassess the way design criteria were set for antilock systems. This reexamination of the problem pro-duced a new set of questions, for which engineers produced new an-swers. Cost and simplicity of operation were eliminated as criteria in most R&D projects as engineers tried to produce designs that would achieve both desiderata.

Chapter 7 looks at the second generation of ABS to make it to the market in the 1970s, produced largely by West German firms. The sec-ond generation of ABS was far more successful both technically and in the marketplace and achieved three skid prevention goals: shortening stopping distances, retaining driver control, and maintaining directional stability. This complex of prerequisites required adaptive control, four-wheel systems, and digital signal processing. These systems were expen-sive but offered clear performance advantages, a status that Daimler-Benz and BMW were able to exploit in their marketing of high-end cars with ABS. The success of the systems that met these performance criteria is also dependent upon improvements in microprocessor production and reliability. Bosch's ABS, introduced to industry accolades in 1978, became the benchmark for subsequent generations of antilock braking. It was also the first production use of microprocessors in the automobile. In the final analysis, engineers had to address performance first, as Bosch did, with little concern for the cost of the system. At the same time the corporate behavior of Bosch and Daimler-Benz can be contrasted with the behavior of American automakers. Bosch and Daimler allowed their employees to play major roles in the knowledge community; far more papers in engineering journals were written by Bosch engineers than by engineers in any other company; Daimler was the second largest producer. In the case of ABS, the two corporations most open to participation in the knowledge community reaped clear benefits from their openness.

Chapter 8 offers an overview of the design of ABS to introduce a model of engineering design knowledge that attends to both the nature of engineering knowledge and the social structures necessary for the production of design knowledge. This chapter also presents an examination of the peculiar function of proprietary knowledge in the knowledge community. One of the strikingly counterintuitive findings of this study is that even with the private sector orientation of the development of ABS, engineers still communicated widely about their work in both formal and informal ways. Despite Derek deSolla Price's claim in the 1960s that engineers do not write, the ABS knowledge community relied on journal articles to communicate and to expand their thriving community. The central role of Bosch and Daimler is best seen in the quantity of their engineers' publications. Chapter 8 takes up this contradiction of public proprietary knowledge and offers a synthetic and more theoretical portrait of design knowledge and the knowledge communities that produce it.

The book ends with an examination of the legacy of ABS. The epilogue was not a part of this study originally, but the need arose in the mid- and late 1990s to explain the unanticipated effects of antilock systems. At that time organizations as diverse as Consumers Union, the U.S. National Highway Transportation Safety Agency, and the Insurance Institute for Highway Safety (IIHS) examined whether ABS was in fact making automobiles safer. The first statistical studies, performed by the IIHS, caused alarm as they seemed to show that ABS-equipped cars in the United States were more likely to be involved in single-car accidents. Wasn't this exactly the kind of event ABS was designed to protect against? Looking at this crisis of confidence in ABS is particularly interesting in the context of the story I tell here because by the mid-1990s the ABS community as it was in the 1970s had dispersed, and so the community looking at the effects of ABS was for the most part independent of those who developed ABS. From the mid-1960s on engineers believed that ABS would have to operate as a black box—an automatic, self-actuated system that would work regardless of the driver's behavior. It was precisely this assumption that led to questions about the efficacy of ABS in the 1990s, as design engineers realized that drivers' behavior was an important factor in whether ABS worked properly and the notion of black-boxing the user was just naïve. This conclusion returned skidding to its original status as a problem of interaction between the car, the driver, and the road and provides yet another cautionary tale about simple technological fixes when pesky, unpredictable humans are involved.

Acknowledgments

▬ ▬ ▬ ▬ ▬ ▬ ▬

Hitting the Brakes is dedicated to the memory of Michael S. Mahoney, a professor of history and the history of science at Princeton University for forty years. On 23 July 2008, Mike unexpectedly passed away. He was my mentor and friend and a strong influence on this work. A master teacher and a thoughtful scholar, Mike took particular pleasure in speaking to many audiences, from software engineers to historians, undergraduates to high school teachers. In part he was able to do this because he scrupulously followed his own curiosity; to Mike, history was about finding the right questions to ask—of practitioners, texts, and artifacts alike. Ideally, interesting questions yielded great stories, such as Christiaan Huygens's disputes with his clockmaker and the tales Mike heard firsthand from Bell Labs engineers involved in the development of UNIX. Mike's questions drove my thinking in this book, as I wondered out loud in his office why so much supposedly proprietary knowledge was sitting in the library at the University of Stuttgart. He replied simply, "That's what you need to figure out." This book is the result of trying to figure that out.

My debt to him is far deeper than my admiration for his scholarship. Mike's death, one year before he planned to retire, cut short a life we all hoped would continue for years, allowing him not only to complete his work on software engineering but, more important, to live out his retirement and see the grandchildren he so cherished grow up. The balance between work and life that Mike valued and achieved is part of what makes his loss so difficult to accept. He did not wait until retirement to live, and he often cautioned me to keep life in perspective: books will wait, but children cannot. He seemed to have figured out how to find joy in life, through his scholarship, through his family and friends, through athletics and his hobbies. He was a consistent voice of reason in an environment that, at times, has tended toward unreasonableness. His work and life continue to inspire me and offer hope that a life well lived is possible in the academy. I present this book to his wife, Jean, as a token of my intellectual and personal debt.

In writing a work about knowledge communities it is impossible not to reflect on one's own intellectual community. Like the engineers I study, my broader intellectual community has made a great difference in the kind of questions I ask in my research. To these people and many others I owe a debt of gratitude. I am often asked how I got interested in antilock brakes. This book began as a dissertation for the Program in the History of Science at Princeton University, where I was seeking a topic that would allow me to investigate the practices of engineers. I was particularly curious about the ways disciplinary arrangements affect the production and movement of knowledge. In a discussion with Harold James about the research and development practices at Robert Bosch, GmbH, I wondered out loud why I couldn't write on something like antilock brakes, which clearly present problems that fall between multiple disciplines, including the electronic control of hydraulic devices and automobile safety. James's response was the no-nonsense answer I had come to expect from him: Why not work on ABS instead of wasting time looking for a topic that was like ABS? So off I went to Germany, where I encountered both unimagined obstacles posed by corporate archives and a treasure trove of very interesting publicly available information in university libraries. Sorting through this material produced three hypotheses: one, not so novel, that engineers produce new knowledge rather than simply apply it; two, more original, that engineers are drawn into problems communally and that dynamic communities form around particular problems that privilege certain practices and solutions; and a third, subversive hypothesis, that regardless of disclosure agreements, engineers in the private sector share their knowledge across corporations. This book aims to explicate these claims.

I have been forced to reflect on my sources because they turned out to be so different from what I initially expected. But this surprise only heightens the importance of sources and especially their repositories. For the historian above all others, those people who tend and provide access to sources from the past, even the very recent past, are indispensable. I must thank several librarians, curators, and archivists at many institutions. I have depended on materials at the following institutions: the University of Stuttgart, the Institution of Mechanical Engineers, Warwick University, Imperial College, the Society for Automotive Engineers, Pennsylvania State University, the University of Michigan, Fordham University, the University of South Carolina, and most of all Princeton University's Firestone and Engineering Libraries. Archivists at Robert

Bosch, GmbH, the Public Record Office in London, and the National Archives in Washington, D.C., helped me find materials without which this book could not have been written. Roman Angerman at Daimler-Benz provided assistance in locating the cover photo. In addition, access to technological *things* played a critical role in the development of this work, and for this I owe special thanks to the Coventry Transport Museum and the Mercedes-Benz Museum. Finally, I must thank the dozens of engineers who spoke with me regarding both antiskid systems and their experiences of engineering communities and knowledge.

Friends and colleagues—often there was no difference between the two—provided support, motivation, and camaraderie to see this project through. Above all, the faculty and students in the Program in the History of Science at Princeton University provided a perfect environment to begin both an education as a historian of science and technology and an obscure project about automobile brakes. I must thank Gerald Geison, Norton Wise, and Michael S. Mahoney for taking a risk on an unknown person with no background in the history of science and admitting me to a wonderful department at Princeton and for continuing their interest in my work and career. The support of these faculty members continued during my years at Princeton. I must also offer my thanks to Harold James, Eugenio Biagini, Mary Henniger-Voss, and Angela Creager for their interest in this project and my education. My fellow students at Princeton have had a huge impact on my intellectual life and have remained friends. Special thanks for an ideal balance between serious scholarly inquiry, compassion, and laughs to Leo Slater, David C. Brock, Eric Brown, and David Aubin. Kara Fulcher was there to welcome me to my first archival research trip to Germany and has been an important friend and confessor ever since. I owe special thanks to the people who read, heard, and offered insightful comments on parts of this work: Terry Shinn, Bernward Joerges, John Krige, David Mindell, Eric Schatzberg, Donald MacKenzie, Jamey Wetmore, Jack Brown, Peter Galison, David Hounshell, Bruce Seely, and Walter Kaiser. Joseph Pitt, Peter Kroes, Pieter Vermaas, Anthonie Meijers, Larry Bucciarelli, and Alfred Nordmann provided an education in engineering epistemology. The anonymous reviewers at Duke University Press provided important comments that have improved this book significantly, as have Duke University Press editors Miriam Angress and Pam Morrison, with special thanks to my copy editor.

I am indebted to the two universities where I have worked since leaving Princeton, the History Department at Fordham University and the departments of history and philosophy at the University of South Carolina. I have been so fortunate in finding colleagues who have engaged my work. At Fordham my friend Nancy Curtin showed interest in a project far beyond her own research, and we shared several perfect days on the Pelham Bay golf course. Michael Latham and Kirsten Swinth provided friendship and advice in my first years of balancing teaching and research. After my move to the University of South Carolina I found equally hospitable history and philosophy departments and a wide interdisciplinary community in science and technology studies. Tom Lekan, Carol Harrison, Kay Edwards, Larry Glickman, Jill Frank, Lacy Ford, Joe November, Ken Clements, and Michael Dickson have been central to the intellectually rich environment in Columbia. Discussions with Sarah Baxter have substantially deepened my understanding of engineering. Joe Brunet suggested the book's title. Davis Baird's role was exceptional; not only did he encourage me to take seriously and take on the philosophers and their conventional epistemologies, but he worked very hard to make possible my move to the University of South Carolina. His own unorthodox work has been an inspiration.

Last, but most important, my family has been supportive of, patient with, and curious about this project and the career I have chosen. My parents, Jim and Elaine Johnson, have shown an unwavering commitment to my education, and I cannot thank them enough for giving me the freedom and confidence to build the life I have today. I am thrilled to share the trials and joys of an academic life with my sister, Katie. You are the kind of scientist we historians of science yearn to study! I owe thanks to my mother-in-law, Daphne Stevens, for her help and support over far more than just the production of this book. You are truly an amazing person. To my husband, Mark Stevens, our life has been an adventure, crossing the continent and full of laughs. It is your understanding of the world around you that inspires me to explain the practices that are not about words, but about things and the imagination. Finally, to Evan: your life has spanned the transformation of this work from dissertation to book. But your transformation from a baby to a real little person with ideas and opinions of your own is far more amazing. Be yourself.

one

Design and the Knowledge Community

In 13 August 2000 an anonymous author with the screen name "Starrion" posted a review of the 2000 Mercury Sable on the website Epinions.com.[1] Starrion told the following story:

I was on Rt 24 headed for Walnut Creek, California. And I'm approaching my exit. It's a beautiful day in the San Francisco area, I've got the sunroof open, going 75, the local radio station playing, and all is well. I know I need to drop a lane, and when an opening appears behind a smart-looking silver Grand Am, I take it. I have only a few feet to spare. Just as I enter the lane, a cloud comes up from the road, the Grand Am's brake lights come on, and she panic-stops. Just as I put my foot to the floor, I see a tire twenty feet in the air—complete with aluminum hub—come bounding over the Pontiac headed right for me. I brake ever harder and the Sable stops flat. No control loss, pointed straight and I out-stopped the Grand Am. The tire lands about a foot from my passenger side door and rebounds into panicking traffic behind me. It was so close that chips of rubber landed in my car through the sunroof. As I started off, I noticed a Jeep Cherokee sans rear-mounted spare about a half a mile up. As I made my exit, I was very glad I had clean underwear in my suitcase.

How was this ordinary family sedan able to avoid this hazard and come to an uneventful stop? Starrion's answer came in the next line of the post: "The Sable LS has great brakes. Better than any other rental car I've driven. This car had ABS or else I wouldn't be writing this. I highly recommend it. Extra highly." While there are many reasons for the differences in the ways various cars handle, in this case, the car's behavior is typical of cars with antilock braking systems, or ABS.[2] When Starrion slammed on the brakes, the Mercury Sable went into an antilock mode. Its computer compared the angular speeds (i.e., the speed of rotation) of all four wheels. The computer also compared the angular velocity of the wheel to the linear velocity of the car. If the computer determined that one or more wheels were decelerating more quickly than the vehicle, the computer could send an electrical signal to the hydraulic system controlling the brakes. That signal would tell the brake caliper (at the wheel that was decelerating too quickly) to pulse, reducing the braking pressure on that wheel. This computerized system is able to apply and release the brakes dozens of times per second, much faster than adrenaline-infused Starrion could pump the brakes. Once the rate of deceleration of that wheel was back in line with the vehicle speed, the brake would revert back to normal operation, even though the wheel sensors would continue to monitor the rate of deceleration for further problems. The result of this active monitoring and modulation of the automobile's brakes meant that the car never started to skid, and as a result Starrion avoided the crash.

Skidding is dangerous for two reasons: a driver cannot steer a skidding car effectively, and a skidding car can take longer to stop. Although it may seem counterintuitive that a car with locked wheels stops less quickly, under most conditions sliding wheels encounter less resistance than rolling wheels.[3] This is also the reason a skidding car often spins; the wheels with the lower sliding resistance move faster than the wheels with the higher rolling resistance. As a result, if the rear wheels start to skid, as they commonly do on a pickup, which usually carries less weight on the rear axle, the vehicle will spin 180 degrees to face the oncoming traffic. No amount of steering correction or countersteering can overcome the friction differential because the direction in which the wheels are pointing does not matter when wheels are sliding, only when they are rolling. Because of the car's ABS, Starrion avoided skidding and therefore maintained directional control, making possible the emergency lane change and quick stop. In this scenario, the proper use of antilock brakes allowed the driver to avoid losing control, to change lanes under emergency braking conditions, and to stop as quickly as possible.

The antilock braking system is a well-designed invention for several reasons. Once invented, it appears obvious. What it does and how it does it are easy to understand, at least superficially. Its function is socially desirable; in principle all drivers prefer not to lose control of their vehicles and to be able to stop quickly. An antilock braking system is fundamentally a technology of control, often represented as a triumph of engineering over dangerous road conditions and unpredictable drivers. In early advertisements, ABS was often presented as a great idea whose time had finally come. Skidding had been a problem on cars for decades. The fact that antilock systems only begin to proliferate in the 1980s points to a difference between ABS as an idea and ABS as a functional, real-world product. From an engineering perspective, ABS is a complex, electronically controlled, mechanical system. At its heart, it is a measuring apparatus, but not one which exists in a clean, temperature-controlled laboratory. Instead, ABS has to send electrical signals along the bottom of automobiles, which vibrate, get wet, dirty, and salty, operate in temperature extremes, and operate at speeds from 0 to over 100 miles per hour. An antilock system has to interface with but not interfere with several other electrical systems. If ABS fails, the braking system itself must continue to work normally. Making ABS meet these real-world conditions proved to be far more difficult than simply dreaming up the idea. A functional ABS required newly invented components, from sensors to computers, circuits to valves, and new understandings of how drivers behave and how other parts of the automobile function. An antilock braking system may be easy to imagine, but it was difficult to make.

The difficulty of the design problem is best shown by the timeline of ABS's development. Patents for antilock systems appeared as early as the 1930s. These early patents were never realized commercially, but they do indicate that some engineers were thinking about inventing devices to prevent skidding nearly a half century before other engineers succeeded in doing so. This lag between the realization of the problem and the appearance of the first technological fix in the mid-1960s is indicative of the complexity of designing and producing antilock systems. Broad awareness of ABS by consumers and market saturation in the United States and Europe did not occur until the late 1980s and early 1990s. In short, developing ABS took most of the twentieth century and involved dozens of different companies and government agencies and hundreds of engineers in Europe, the United States, Canada, and Japan.

THE DESIGN QUESTION: PUTTING IDEAS AND THINGS INTO PLAY

This is not principally a book about automobiles, but about engineering design. I focus on the design of ABS as a case study to illuminate the complicated process of actually designing something. If one draws from recent approaches in science and technology studies and focuses on the practices of engineers—that is, what they do—then the central problem of design is not where ideas come from but how ideas become things.[4] It is worth noting at the outset that ideas that do not work are without value in the marketplace. In the world of engineering, artifacts clearly trump ideas and propositions.[5] This is not to underestimate the role of knowledge in engineering. Quite the contrary, to transform an idea into a functional thing requires knowledge. Furthermore, the transformation from idea into thing or one thing into another usually requires new knowledge, not simply repackaged or applied scientific knowledge.[6] In fact, making substantially new things requires substantially new knowledge; ideas, therefore, do matter.

Knowledge is produced in and by communities. In this book I use a framework I call *knowledge communities*. Through the case study of antilock braking systems I show that knowledge communities are the basic locus of knowledge production in design engineering and much science (and certainly they appear and function in many other human endeavors). A knowledge community is a socio-epistemological structure, and I do not privilege either the social or the epistemological dimension. As a result, I do not want to describe a body of knowledge independently of a community, implying that free-floating ideas and artifacts came together and attracted a community of practitioners. In this scenario a basic causal problem arises: What agents determined the central body of knowledge in the first place? Conversely the community cannot be prioritized either, since it has to be clear what attractor (e.g., a problem or a set of ideas) brought together a multidisciplinary group of practitioners in the first place. Consequently, an analysis of knowledge communities requires a real commitment to symmetry between knowledge and social organization; it may even require the collapse of both into a single entity.

In terms of methodology, knowledge communities lend themselves to historical analysis, or at least historical sociology, because describing the development of a particular knowledge community keeps social and epistemological factors properly integrated. One can track the changing definition of the community's central problem, knowledge, and artifacts while simultaneously watching participants come into and out of the community. In other words, the development of the knowledge community must be described in order to determine whether a knowledge community exists at all. In addition, a developmental narrative shows the dynamics of the knowledge community. Dealing separately with knowledge and then social arrangements creates the impression that both are static, as the epistemological narrative must be held up as the social narrative catches up, or vice versa. The shifts in what knowledge was valued by the knowledge community and the concurrent changes in community membership are very telling in the case of ABS. In fact, these shifts show the contingent process of the social construction of knowledge; ABS had many different possible paths of development, but certain paths were more highly valued than others. There were also important physical and economic constraints on the system's development. To explain the convergence of the ABS design around an electronically controlled hydraulic system produced by Robert Bosch, GmbH, a West German company, in the late 1970s, I must also explain what kinds of knowledge and practitioners were valued. The knowledge community dictated these preferences and values.

There are several other salient features of the knowledge community. Knowledge communities in engineering initially form around a communally defined problem, which I call the attractor. But the problem that sits in the center of the community and that attracts the often diverse practitioners is never static; the community is constantly redefining it, ruling certain solutions and definitions in and out. This is an informal process; in fact, formalizing the process often marks the transformation of the knowledge community into something else, perhaps a professional community. Because the focus of a knowledge community is so narrow (far narrower than the focus of a discipline or a professional society), the group is quite small, varying significantly over time but always remaining intimate. Intimacy in a knowledge community means that the practitioners all know each other, or each other's work. In the development of ABS no more than two degrees of separation existed between members of the knowledge community. As a result of that requirement, knowledge communities rarely exceed more than a few hundred people. Those that grow larger begin to splinter as the practitioners develop discrete problems on which to focus. Furthermore, knowledge communities are nonexclusive, so engineers may work within several different knowledge communities. Because a knowledge community centers on a problem, one can be engaged in several different problems simultaneously. New ideas and tools move into the community in part because participants move between communities, and ideas require human vectors.

Thus I can trace knowledge and community evolution through a series of *how* questions that united the effort to design antilock systems. The nascent community initially asked, How can we reduce skidding accidents in passenger cars? This question splintered into questions about driver education and psychology, tire design, road surface design, and finally brake design. The group that defined skidding as a braking problem continually refined their central question. They first asked, How can we design a braking system to prevent skidding in panic-stop situations without requiring the driver to change his or her habits? Then they asked, How can we determine that a car is about to skid so that a system can prevent skidding? How should we design an electronic system to measure imminent skidding, then a hydraulic system to react to that electrical signal? And finally they asked, How can we make a mass-producible electronic control system that will integrate with the braking system and prevent the car from skidding without the driver even knowing it is there? Whether the systems completely accommodate these requirements is still unclear; the question of assessment is the subject of this book's epilogue. But I want to emphasize the changing questions the community asked and show how these changes were interwoven with the changing membership of the knowledge community.[7]

This project of connecting engineering design to the production of engineering knowledge is hardly new or novel; any appearance of novelty comes from the fact that, compared to scientific knowledge (or science more generally), engineering knowledge (or engineering more generally) is significantly understudied. In the small but growing field of engineering studies, one significant focus has been on design and on the nature of engineering knowledge. In *What Engineers Know and How They Know It*, Walter Vincenti equates design and knowledge by explaining that design and problem solving are synonymous at the working level. He writes, "Day-to-day design practice not only uses engineering knowledge, it also contributes to it."[8] Designing is problem solving, and solutions and designs both constitute new knowledge. But the voluminous literature on scientific knowledge can actually interfere with an understanding of engineering knowledge. Philosophy of science, along with epistemology, the philosophical study of knowledge, makes certain claims about the nature of knowledge, and these claims tend to be incommensurable with notions of engineering knowledge that focus on the process of design. Epistemologists have long focused on propositional knowledge and argued that knowledge is "justified true belief."[9] As a result of these two deeply held principles, philosophers of science have tended to focus on theories as exemplars of knowledge.[10] These analyses of knowledge as universal and propositional do not fit well into technological, engineering, or design knowledge; one needs a broader, less formal, more inclusive concept of knowledge. Engineering knowledge rarely takes the form of formal scientific theories; more often engineers' knowledge resides in the artifacts they design or in the processes of designing those artifacts.

The philosopher Davis Baird has tackled this problem of broadening a definition of knowledge in his book *Thing Knowledge: A Philosophy of Scientific Instruments*, published in 2004. Baird shows that in addition to propositional, theoretical knowledge, science and technology also both need and produce knowledge in physical, artifactual forms. He breaks this concept of the physical instantiation of knowledge into three kinds:

1. Working knowledge: knowledge communicated by the operation of a device or instrument; knowledge by action.
2. Model knowledge: knowledge by representation, using physical models.
3. Encapsulating knowledge: knowledge which both represents and acts; incorporates both of the previous categories and in most cases provides measurements. The best examples are scientific instruments.[11]

The antilock knowledge community actively and enthusiastically produced all of these forms of knowledge. But even Baird's extension of the notion of knowledge is insufficient, because process knowledge (i.e., nonpropositional answers to *how* questions) is also produced in engineering design.

I argue that a broad knowledge category, which I call *design knowledge*, is ultimately the only way to describe the knowledge products of engineering design. Design knowledge incorporates theories and propositional knowledge, Baird's "thing knowledge," tacit knowledge, and skill, but also hybrids of these categories. These hybrids are often called *know-how*, which includes both manual and mental ways of knowing.[12] Design knowledge requires special attention to objects and the ways artifacts both bear knowledge and can constrain and shape knowledge.

Yet parsing technology into "ideas" and "things" without emphasizing the inseparable nature of the two in engineering practice makes no sense in the context of engineering design; engineering design ideas have value only as they relate to real artifacts. On the other hand, artifacts and ideas clearly do not belong in the same ontological category, and as Baird's analysis reminds us, artifacts have been utterly ignored in epistemology. Looking at design processes probes and problematizes knowledge precisely because engineering culture privileges making artifacts rather than proposing ideas. The contrast between making and proposing is as important as that between artifacts and ideas.

Given the complex relationship between actors, artifacts, and knowledge, expressed most directly in the design process, how can design, as the central intellectual activity in engineering, be analyzed? How is something designed? How do engineers and others work together in the process of designing? Design is clearly not a solitary, creative activity that takes place privately in the mind of a designer. Design engineers engage in a social process involving multiple forms of formal and informal communication, verbal, mathematical, and visual. In his book *Designing Engineers*, the mechanical engineer Louis L. Bucciarelli argues that design must be understood as a social activity, a claim that is unquestionably true. He writes:

In the contemporary engineering firm, designing engages more than a lone engineer at a drafting board or work station. The design [of a system] engages a wide variety of people within the firm: research scientist, marketing chief, lab technician, systems engineer, project manager, production engineer, purchasing agent, inventory controller. All can and do influence the design, and all must come to agreement in order to realize the design. The process is thus social, the business of a subculture. Not surprisingly, participants' visions of the social process of designing are strongly influenced by their understanding of the way the things they are designing work. To participants in design, the object serves as a kind of icon that embodies a set of attitudes and ways of thinking that are peculiar to engineering.[13]

Bucciarelli proceeds to develop three case studies of the social process of designing artifacts for commercial production. Most important, he argues that in the process of designing something, the question "What is it?" intertwines completely with the question "How does it work?"[14]

While the question "What is it?" has overtones of aesthetics and marketing, the question "How does it work?" appears at first glance to be more grounded in engineering and technical know-how. One can image a simple scenario in which one individual imagines an object and another figures out how to make it, thus inventing the way it will work. In fact, the common model of invention in the corporation often follows this erroneous assumption. Business models of the corporation attribute the decision of what the engineers will produce to the executive rather than the technical level. Imagine the following scene. The executive says to the engineer, "We want to introduce a machine that will prevent a car from skidding. It should cost less than 10 percent of the total cost of the car and be utterly reliable. We must beat our competitors to the market in order to fully capitalize on this opportunity." The engineers get to work producing a feasible design under physical, economic, and temporal constraints. This schematic view of product design implies that the impetus to introduce this product emanates from the penthouse, while the actual design of the product winds its way up from the laboratory. This scenario seems believable because the ethos of corporate culture states that engineers in the private sector are told what to design. Engineers even participate in this myth by taking responsibility for how a product works but attributing the decision of what it is to the executives. Yet in designing the product the engineers also define what it is, even if the executives (or the nature and sector of the corporation) constrain the possible things they might be employed to design. In *Engineering and the Mind's Eye*, a study of engineering design, Eugene Ferguson claims, "Many engineers deny their influence, insisting that they merely carry out the orders of others. . . . Yet it is in fact the engineers who draw up the shopping lists."[15]

In fact, when an executive decides to allocate human and financial resources to produce a device to prevent skidding, this action is trivial compared to what follows. In product design, ideas, social realities, and physical realities collide. Telling the engineers to invent such-and-such a machine is entirely different from the process of invention.[16] To use a ridiculous example, if an antigravity, time-travel, or perpetual-motion machine could be produced it certainly would sell. But defining the effect the machine has is not at all the same thing as saying what it is and certainly not how it would work.[17] Transforming the vague idea of a device to eliminate gravity, or perhaps more feasibly to prevent skidding, into something functional requires a substantial effort. In the case of antilock braking systems to prevent skidding, this process took some twenty-five years of concerted effort by a defined community of practitioners, from the definition of the problem of skidding in the mid-1950s to a functional product around which market closure occurred in the late 1970s. The route from having a problem to solve—in this case, preventing skidding—and having a device to produce and market is long, indirect, and unpredictable, involving several redefinitions and refinements of the problem itself. To derive some abstract idea of a product from the artifact itself violates Bucciarelli's central argument: what the artifact is and how it works, at any level other than the most trivial, cannot be made independent.[18]

While executive decisions do play a role in allocating resources to R&D projects, engineers control the process of moving from suggestion to prototype. What engineers do in order to produce these functional machines and systems is complicated, messy, and rarely linear or straightforward. In short, the process of design does not consist of discrete multistage activities, where first the executive tells the engineer what to invent, then the engineer imagines such an object, proceeds to work out the details of materials and geometrical arrangements, and then mathematically predicts its performance. In their textbook *Engineering Design*, Gerhard Pahl and Wolfgang Beitz present several flowcharts of design processes (figure 1).[19] But such linearizations are inherently idealistic and misleading.

Eugene Ferguson argues strongly against this neat sequential interpretation of engineering design, writing, "Block diagrams imply division of design into discrete segments, each of which can be 'processed' before one turns to the next. Although many designers believe that design should work this way, even if it doesn't, it is clear that any orderly pattern is quite unlike the usual chaotic growth of a design." Ferguson even claims that the "heart of a design" may be latent in an engineer's mind before a need has even been expressed, emphasizing the raw creativity of engineering design rather than its responsiveness to corporate needs.[20] Thus a schematic view belies the unique and entangled nature of engineering design.

In schematic overviews of engineering practice, the process of engineering design is often split into two broad activities, which are linked by a feedback cycle. A design is first conceived, then analyzed. By "analysis" I mean the process of breaking down the design to consider

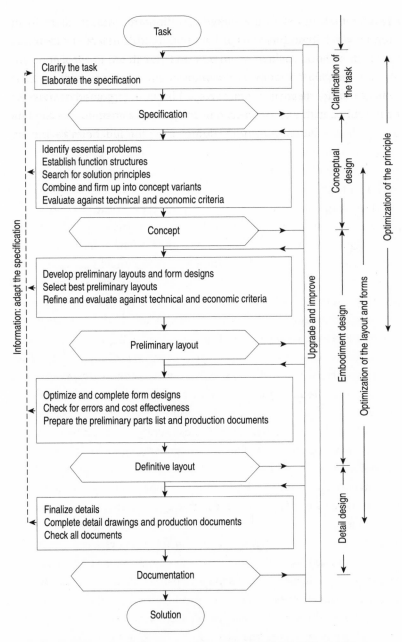

1 Steps of the design process. Reprinted from Gerhard Pahl and Wolfgang Beitz, *Engineering Design: A Systematic Approach* (New York: Springer-Verlag, 1988), 41.

the operation and structure of each of the parts of each component individually. These questions can be mechanical, structural, chemical, electrical, or social. What are the forces at play in the phenomenon the device is supposed to produce? How will they affect, constrain, or determine the structure of the artifact? How does structure affect the device's operation? Can the artifact be produced? Can it be mass-produced? Will drivers' habits be able to accommodate any new practices the artifact requires? Broadly, any analysis must answer two questions: Will the components function individually? and Will the system perform reliably in real-world, real-time situations? Analysis can be done in an engineer's mind, on paper, on a computer, or in prototype. If the analysis shows less than optimal performance, the design can be reconceived. Viewing the design process as a feedback cycle rather than a linear process is more accurate, yet feedback still implies that the mental conceptualization of a device and the physical testing and analysis of models and prototypes are discrete operations.[21] In practice, testing and analysis play integral roles in the ways ideas are generated, and testing should not be perceived simply as an effort after the "real" creativity has occurred. Some kind of assessment, whether formal or informal, must be performed in order to make an initial design. In other words, even initial design choices require analysis. Engineers must consider what material to use, what dimensions the prototype should have, whether the signal path is feasible, and so on. There is no discrete feedback here; assessment and even partial testing are part of generating the prototype.

Engineering design textbooks often describe a simplified feedback cycle of conception and analysis as the "method of engineers," similar to an abstract concept of a "scientific method," wherein hypothesis formation precedes empirical testing, which leads to theory creation.[22] However, the separation of conceptualization from analysis completely ignores the actual practice of engineers, for whom these activities are not so easily distinguished. The analysis is part of conceiving the device, method, or idea; when working to produce new products, the initial conception of the device is inseparable from a discussion of how it will work. Analysis produces the answers to the questions "How does it work?" and "What is it?" One of Bucciarelli's examples is illustrative of this problem. Tell an engineer to design a phone system.[23] The first iteration of questions will focus on what a phone system is and what its components are. But Bucciarelli then asks what it would have meant to conceive of a system that had never before existed. This assumption changes the nature of the question "What is a phone system?" from something pat to something profound. You may be able to explain the idea of what a phone system does from a user's point of view, but this will not tell anyone *what it is*. The idea of the phone system cannot be split from the reality of the phone system, particularly a priori. Without some version of a telephone system, there can be no idea of a phone system.[24] Object cannot be extracted from idea, theory from mathematical model, or form from function.

KNOWLEDGE, COMMUNITIES, AND ARTIFACTS: A CASE STUDY

If knowledge is inseparable from an artifact, what does it mean to produce knowledge in the context of design? There is the knowledge that frames the initial steps of artifact design, the knowledge produced in the process of designing the artifact, and the knowledge needed for the artifact's use. The artifact does more than simply embody the knowledge used to produce it. Baird's articulation of "thing knowledge" shows this view to be far too limited; artifacts bear knowledge in many different ways, not simply as instantiations or embodiments of designs. Knowledge is not only embodied by the object but is also produced by designing it. In addition, both user and producer possess and wield legitimate knowledge.[25] Ferguson refers to the concept of "deep knowledge" to explain the level of knowledge a designer possesses about a product as a consequence of having designed it.[26] Vincenti borrows the concept of the "operational principle" from Michael Polanyi to describe knowledge produced in the process of design.[27] To define the operational principle Polanyi writes, "A patent formulates the operational principle of a machine by specifying how its characteristic parts—its organs—fulfill their special function in combining to an overall operation which achieves the purpose of the machine."[28] Thus the production of knowledge in the design of artifacts constitutes a complex process with several different forms of knowledge as outcomes. This complex knowledge system has not been fully explored in the history, philosophy, or social studies of technology.

While design knowledge and knowledge communities can be studied in the abstract, far more useful explanations emerge from case studies of product and process design. Admittedly, case studies have a problematic relationship with generalizable knowledge, but they enable one to examine the concrete connections between thing and idea in real time and in the social atmosphere in which knowledge is produced.[29] Historical case studies in particular allow one to focus on the relationships between (1) knowledge: the generation and evolution of understandings of both phenomena (e.g., skidding) and artifacts (e.g., high-speed valves); (2) artifacts: the development of prototypes and their evolution into marketable, mass-producible products; and (3) communities: the changing social structure and composition of the knowledge community. The development of ABS for passenger cars is a particularly useful case study because it provides an opportunity to examine further the problem of engineering design as it occurred in both private and public sector settings. Invention presents an array of common difficulties faced by engineers in research and development: the problems of working between disciplines; of creating a marketable, affordable, and reliable product; of generating new mathematical models and embedding them into electronic processors; and of forming a multidisciplinary, multilingual community that spanned institutional settings and corporate competition. The development of antilock systems serves to demonstrate some of the tactics engineers use to construct common knowledge in the face of the need for secrecy in product development. In addition, antilock systems did not develop predictably; the process was contingent and could not have been mapped out ahead of time, a dimension that is particularly interesting as problem-driven models of innovation and targeted research initiatives gain more currency.[30] As a result, antilock systems are an ideal case study for trying to understand the dynamics of product design more generally, precisely because so many complex factors that commonly arise in design or invention and innovation are present. In *The Reflective Practitioner*, Donald Schon emphasizes that the way a question is framed is critical to its solution.[31] One must ask, "What was this particular system designed to do?" Sometimes design criteria can be read from the final design, but often the engineers have documented those criteria for their peers in the publications surrounding the introduction of a new design. Accounts of design in process are often different from accounts taken from the finished work, as the finished work often provides a teleological version of its history and design goals by excising all dead ends and previous versions of design criteria. The design specifications, as well as the arguments defending these choices, are important windows on the process of engineering design. To phrase it in Schon's terms, if a given antilock system was the answer, then what were the questions? Questions and answers had to be worked out together.

While the term *design knowledge* does include knowledge of a verbally incommunicable nature, called *tacit knowledge*, it also must include knowledge of a highly communicable sort that is often left unsaid (some might call this implicit, as opposed to explicit, knowledge).[32] Building a prototype clearly yielded knowledge of a hands-on or know-how sort. Designing instruments for measuring phenomena also generated knowledge of a more abstract nature about how the engineers wanted the measuring device to perform. If a given instrument was the answer, what was the question? This knowledge could have taken the form "We wanted the machine to measure factor X, and either it produced the results desired or it produced something else." "Something else" could have referred to inaccurate or wrong results, that is, a design that did not work properly, a failure. Alternatively, the unexpected results may have reflected back on the underlying concept of the problem. Engineers often unknowingly asked the wrong questions; figuring out the right questions to ask was part of the generation of new knowledge.

Machines developed to detect that a car was about to skid often produced unexpected results. Some engineers thought of these machines as examples of failed designs, but others took the unexpected results and revisited the machine's design to try to understand why such measurements were being produced, asking whether they should modify their expectations or the instrument's design. In other words, they rethought the question. Did the machine provide an unexpected result because the fundamental idealizations did not reflect the way the real-world system worked? Many engineers have remarked that they learned more from designs that failed to produce the desired results; that is, more knowledge was generated by the failed design.[33] In their study of the design of the Britannia Bridge, Nathan Rosenberg and Walter Vincenti point out that design engineers can generate important valuable new knowledge even when they do not produce a valuable or functional object.[34] All of these different kinds of knowledge—a successful design that works as predicted, an inaccurate design, or a design which produces unexpected results—as well as the tacit knowledge gained by the process of making the machine are integral to the development of the end product. In their textbook *Engineering Design,* Pahl and Beitz refer to design as "the most complex form of the learning process."[35] By building machines one learns how to design better machines, and this knowledge is often communicable, and frequently published or otherwise made accessible, to engineers working on similar problems.

In the case of antiskid devices, the knowledge was produced and diffused by the diverse community of engineers who produced the objects. Looking only at the most successful electronically controlled, antilock braking system that Daimler-Benz and Robert Bosch GmbH introduced in 1978 does not tell the history of all antilock braking systems. A linear history of who made what inside Daimler and Bosch does not lay open the engineering culture and communities that made the production of ABS possible. A vast field of new knowledge had to be produced in order to make a functional and economically feasible antilock braking system. Engineers produced this new knowledge through the process of designing both concrete objects and abstract models and theoretical tools. Often, but certainly not always, intermediate design steps can be read in a final design. Steps which were, in themselves, failures were often effaced from the apparent genealogy of the machine. Yet the knowledge produced by failed design was developmentally critical, as it led to new ways of thinking about the design of antiskid devices.

Reading a machine's history from its final form often produces a distorted linear picture of its development, excising the interesting dead ends that nevertheless produced critical knowledge. It is also important to avoid an evolutionary, or tree, structure, because tree structures explain diversity from a single origin. Compared to engineering design, tree diagrams depict information moving the wrong way; design does not require explaining diversity from a single origin, but rather convergence from multiple origins. In the case of antilock systems, theoretical tools and machines from brake design, vehicle dynamics, tire design, microprocessor design, and avionics (and more) came together in the final prototypical design of an antilock device. From a marketing as well as engineering point of view, why did this device achieve closure and erase its less successful competitors?[36]

Much of the knowledge integral to the production of a feasible antilock system came from the development of devices to measure skidding for the purpose of improving road and tire surfaces; in this sense ABS is a particular case of the development of what Terry Shinn calls "research-technologies."[37] Engineers produced these machines under completely different constraints and resources with different goals in mind, yet their existence shaped the way antilock system designs evolved. Henry Petroski has written two books on the value of failure in engineering; he points out that what looks at first glance like a failure or dead end may not be.[38] However, looking for the ancestors of a device will not show all the failures and indirect influences that led to the series of designs produced by engineers. The reason for this is simple: unearthing the origins of all ideas and influences from different projects requires a complete overview of the process that no one individual possesses. Therefore, while artifacts are important and often overlooked historical sources they must remain attached to the humans who made them in order to explain the transfer of ideas into and out of knowledge communities. Ideas, which may be communicated through devices or pictures rather than words, are still dependent on people for their circulation. This is one way to interpret Bucciarelli's insistence on the social nature of engineering design.

Consequently, one must look at engineering *projects* in order to investigate the social dynamics of engineering design. There are at least two important social dimensions. First, looking at the members of a team working on a single project within a single laboratory or corporation reveals a complicated network of information exchange, prioritization, and, most obviously, power relationships. In *Soul of a New Machine*, Tracy Kidder examines these local politics at play in the development of the Eagle minicomputer in the incubator of Data General. In Kidder's account the only episode of the story that takes place outside of the two campuses of Data General is an attempt to examine a competitor's machine for the purpose of reverse-engineering it.[39] While the examination of the internal dynamics of a single company well serves Kidder's narrative, and the development of minicomputers took place in a highly competitive and secretive environment, no engineering project exists solely in the vacuum of a single company. Even in the case of Data General, the actions of its competitor, Digital Equipment Corporation, shaped and helped define the path that Data General's executives and engineers took. Therefore, the second social dimension that informs product development comes from outside a single laboratory.[40] It would be nearly impossible to understand the origin and evolution of both ideas and things without some appreciation of the influences from outside the team. These influences may be from the engineers working at competing companies, from engineers working on other projects at the same company, or from developments coming in from academia or government agencies and laboratories. They may also be the social and economic realities of the society in which products are developed, which govern acceptable costs and other elements of production. But unexpected outside influences also play important roles. In one episode in the development of ABS, an engineer turned on the radio in his car and experienced a eureka moment in which he realized that a radio tower near a test site was possibly causing interference with the electronic signal of the braking system's control unit. Engineering ideas and insights also come from trivial everyday experiences, none of which is captured in flowcharts. Naturally, the changing society in which engineers are working also affects the design process. An understanding of the development of a single product must take into account this larger group of influences. Even so, our understanding of the process will never be complete.

THE STRUCTURE AND DYNAMICS OF THE ABS KNOWLEDGE COMMUNITY

The complicated history of antilock braking systems also serves as a case study for understanding knowledge communities, which provide a framework useful for discussing technological developments. The Elmer Sperry Award for "advancing the art of transportation" was awarded in 1993 to three engineers from Robert Bosch GmbH, for their development of the Bosch electronic ABS.[41] Heinz Leiber, Wolf-Dieter Jonner, and Hans Jürgen Gerstenmeier developed Bosch's successful ABS at Teldix, GmbH, then continued their work between 1973 and 1975 when the ABS project was moved to Bosch's main research and development laboratories in the Stuttgart suburb of Schwieberdingen. However, these engineers were only three in a community of a few hundred people who were involved in developing the ideas and technologies which underlay the system patented and produced by Bosch. While the work of Jonner, Gerstenmeier, and Leiber certainly deserved credit, placing credit for this invention in the hands of three individuals without mention of the community misrepresented the production of knowledge in the private sector. On the other hand, the Sperry Award was a public acknowledgment by practitioners that these three engineers were recognized as the leaders of their community.

Harry Collins, in a sociological study of induction and replication in science titled *Changing Order*, provides a means for analyzing the development of expertise in science.[42] Collins sets the locus of knowledge in the community of practitioners. Knowledge is acquired through direct and personal contact with the community. Social organization becomes central in the process of discipline building, particularly the professionally sponsored conferences where a group of researchers can establish boundaries between themselves and other communities over time. Experts emerge from this community. The establishment of a self-recognized subdiscipline or research field is required before experts can be acknowledged. Collins's emphasis on community and tacit knowledge is particularly useful when his model is applied to technological development.

The emergence of ABS experts in the 1970s marks the maturation of an antiskid community and coincides with the market introduction of several antiskid devices. From 1958, when the first international meeting of the nascent skidding community was held in Virginia, to 1978, when the first generation of successful electronic antilock braking systems was introduced, the development of antilock systems was always a multidisciplinary problem, with specialists coming from precision mechanics, hydraulics, electrical and electronics engineering, instrumentation and testing, and physics. In addition, skidding research did not fit into traditional company departments either. Most companies interested in designing and selling an antiskid device had to reorganize or create a new division for development of ABS. Yet the human and instrumental resources available at different companies were dependent on the company's other specialties. The playing field was clearly not level, though in 1958 it was not clear which way the field tilted.

Initially, experts were limited to their particular discipline of previous training; for example, the electronics engineer on the ABS team remained an electronics specialist. The notion of an ABS specialist did not yet apply; consequently, ABS expertise often resided in the team. In addition, no single company dominated ABS; no fewer than two dozen companies worked on antiskid system design, and even more research was performed by government agencies, industrial coalitions, and university research centers. Establishing expertise in the field required grabbing the attention of the whole community, making expertise a publicly visible attribute. The complexity of the ABS knowledge community, with its multilingual, multidisciplinary, private and public sector participants, heightened the importance of experts and made conferring expert status an important function of the mature knowledge community. Companies valued employees who were named experts, since they saw a relationship between the praise of their engineers and success in the marketplace.[43] However, being an expert on a multidisciplinary problem like skidding should not be oversimplified; the problem of expertise involved both the internal and external dynamics of the knowledge community. Experts were experts in part because of the knowledge they brought to the problem at hand, knowledge often generated outside the particular knowledge community. Through this route, expert knowledge and status could be carried over from one knowledge community (say, electronic fuel injection) to another (ABS).

However, knowledge communities did not define engineers' identities in the way a specific field might define one's professional identity. An engineer might self-identify as an electric or mechanical engineer, but also be part of several knowledge communities. Heinz Leiber, one of the Sperry Award winners, is an example of such a multiple community participant. Leiber was by training a mechanical engineer, working on high-speed valves for an avionics company, designing navigational systems. He began to engage in the development of ABS because of Teldix's development of a high-pressure, high-speed valve. Thus he participated in at least three knowledge communities: one focused on avionics and aircraft navigation, one centering on high-end small hydraulic valve development, and one working on skid prevention. As the skidding community redefined its focus on electronic control in the 1970s, Leiber reoriented along with the community and brought more of his avionics experience to bear. But his professional identity was based on an amalgamation of these communities rather than a single point. His highest achievement may have been in the antilock field, but this bestowing of expert status also was an external recognition rather than a personal statement by Leiber.

The means for achieving expert status were constructed as the community was defined. The ability to invent solutions to widely acknowledged problems was one criterion for establishing oneself as an expert;

this often correlated with patent activity. However, even earlier, engineers who introduced theoretical tools that were general enough for designers of different systems to use emerged as the primary experts of the skidding community. These new knowledge producers fell into two primary categories: people interested in modeling vehicle dynamics, many of whom worked for universities or industry research groups, and engineers, often in the private sector, who invented and standardized testing technologies. These engineers had a particular stake in making their ideas public. It is important to note that the theory builders came from both public and private sector institutions and were not seen as a subcommunity separate from the device inventors. Surprisingly, government and university employees received a proportionate share of the patents. The members of the skidding research community erased the hard line between the kind of work done by engineers in the private sector and the kind done by engineers in the public sector.

Engineers who developed the instruments needed to produce measurements of the phenomena of skidding relied far less on published articles and far more on the informal and face-to-face exchange of skills. There are several reasons for this peculiarity, reflecting the nature of the knowledge being shared. Following Baird's claims about "thing knowledge," the designers of measuring instruments were primarily producers of what Baird refers to as "encapsulating knowledge," a kind of thing knowledge that both represents and produces a phenomenon.[44] For these practitioners the specialized instruments they developed bore their knowledge far more effectively and completely than articles about the instruments could. This turn away from publishing reflected the nature of the knowledge being shared rather than concerns about the proprietary nature of the work, since metrology was probably one of the least proprietary fields in antiskid research.[45] Because the performance of similar products would be compared by potential customers, engineers from different companies collaborated on testing methods and performance standards. Naturally, each company had a particular approach to setting standards that favored the kinds of product designs that company generated. Still, the collective establishment of testing protocols was generally useful for companies, and engineers encountered little resistance from their employers in sharing this kind of information. In the 1970s all manufacturers of antiskid devices faced similar skepticism from automobile producers, so presenting a coherent set of standards aided the cause of antilock braking in general, not just individual companies' positions in this new industry. The sites for the exchange of testing information were conferences and shared testing sites.[46]

The notion of a knowledge community does important work in establishing the milieu in which engineering knowledge is produced. The co-construction of knowledge and community, built in the context of designing a new product for the consumer market, is a key dynamic and must be explored as a first step to understand the dynamics of research and development. Linear flowcharts of the design process, even when feedback loops are embedded in them, oversimplify the process because they fail to account for the input of ideas from outside a particular design team. Although scholars of engineering studies need to track the development of an idea within a single design team, finding and explaining the human vectors of new ideas is just as important. Frameworks must be found that help answer the questions: how do ideas get into (and out of) the flowchart? Of the infinite variety of ideas circulating throughout engineering at any given moment, why do certain ideas permeate the design process and not others? Why does a certain engineering problem attract attention at a particular moment? The knowledge community is a framework designed to help answer these questions. Furthermore, the knowledge community is a device that is aimed at seeing mesoscale interactions—focusing neither on institutions nor on individuals, but rather foregrounding processes and dynamics to answer the question of *how* knowledge is produced, used, and moved. Most important, the framework presented here intentionally examines knowledge and the social structures of knowledge production in their naturally entangled state.

two

A Genealogy of Knowledge
Communities and Their Artifacts

If one takes seriously the knowledge community as an organizational principle for understanding the development of a new technology, then the story detailing that technological development must pay joint and simultaneous attention to the social dimensions of community formation and how new knowledge is being produced. In the case of ABS, it is most useful to consider the artifacts engineers were producing as bearers of that knowledge.[1] Antilock braking systems have dual genealogies; one family tree traces changing social arrangements and the other follows the development of artifacts. The coherence of these two genealogies is one piece of evidence for the existence of a knowledge community. These genealogies also help to periodize the development of the technology; as membership in the community changed, so did the kinds of artifacts members produced, and as artifacts evolved, so did the kinds of practitioners who were interested in them. In the end, the trick is to unite these two lines to see the development of a knowledge community with its inseparable social and epistemological dimensions.

Focusing on the artifact as the locus of knowledge complicates the automobile industry's chief consumer outreach strategies regarding ABS. These advertising strategies present ABS as a particular kind of idea, and that presentation embeds the industry's own ostensible motivation for making ABS into a myth about its production. Most consumers in Europe and the United States became aware of ABS in the early 1990s, when the acronym ABS started to appear on the back of automobiles equipped with antilock braking systems. This simple advertising gimmick gave the appearance that ABS was bursting onto the scene, instead of the reality that after decades it was slowly becoming cheap enough to permeate the family sedan market. The apparently sudden appearance of ABS effaced not only the people who created it but also the difficult process of developing it. Antilock brakes were offered as evidence that the automobile industry cared about the safety of drivers, not as a negotiated series of solutions to initially intractable technical problems involving the reliability of electronics, the mechanical complexity of high-speed valves, and the construction of algorithms that could compare angular and linear velocities. To be sure, engineers did want driving to be safer, but this had little impact on the kinds of antilock systems they developed. Antilock braking systems did not burst on the scene; they were laboriously crafted and carefully escorted to a market that, in many ways, had resisted both gradual improvements and the entry of electronics.[2] Focusing on the genealogy of ABS as an artifact reintroduces a series of otherwise effaced systems and human actors back into the narrative of ABS development, in part because it does not limit the ancestors of ABS to devices that were market successes. In fact, the evolutionary dead ends and failures were critically important to the development of knowledge about skidding, and their inventors were active participants in the final form of ABS. Rescuing the devices also rescues the forgotten engineers, which further reinforces the point that the devices were bearers of knowledge and makes a tacit argument for the inseparability of knowledge and the community.

However, just as in natural evolution, there can be apparent relationships that prove to be false; artifacts may look similar and communities ought to intersect, but these appearances and apparent relationships can be misleading. Often these connections are misread in an effort to produce a longer term history than a technology actually warrants. Such history making is quite common in politics; witness, for example, efforts to invent national and cultural traditions in the nineteenth century.[3] Perhaps the best known example of such technology history making is the effort to trace the history of computing back to the mechanical calculators of Pascal and Leibniz, which share nothing either operationally or epistemologically with the electronic computer. Nor can a continuous community interested in computation be traced back to the seventeenth century. Yet such artificially long histories are commonplace in introductions to computer science. The antilock braking system has similarly artificial roots stretching back to the 1930s, when both the British and German governments approved patents for "skid preventing" devices.[4] The problem of skidding does have a history as long as the automobile's, but it is wrong to assert a genealogical connection between the antilock devices of the 1970s and designs from the 1930s. The successful electronic systems of the 1970s were not the descendants of the mechanical systems developed prior to World War II, and a community oriented around the problem of skid prevention devices has not been in continuous existence since the 1930s.[5] Not one of the devices patented before the war was ever produced, and the designs did not provide a foundation for postwar designs. While Bosch and other companies claim that these prewar patents establish their priority in the field, these designs have virtually nothing to do with the systems that were first marketed in the 1970s.

Instead, by examining actual artifacts and their institutional and individual inventors and users, we can see that the history of ABS developed in four fluid and continuous phases: (1) a knowledge community of public health practitioners and transportation policymakers that generated initial questions about skidding; (2) a second knowledge community that responded to those questions by developing a series of testing instruments to measure the phenomenon of skidding; (3) a third knowledge community with large representation from the private sector that actually developed antilock devices; and finally (4) a fourth knowledge community that struggled in a highly competitive market to maintain its existence in the face of pressure for proprietary intellectual property.

PHASE I: AUTOMOBILES AND PUBLIC HEALTH IN
THE 1950S AND 1960S

Skidding came to engineers' attention in the 1950s through investigations into the safety of automobiles and roads in postwar society. Responding to concerns about what the American politician Daniel Patrick Moynihan called the "epidemic on the highways" in 1959, government agencies and academics in Western Europe and the United States began a series of statistical and epidemiological studies of the risks of being on the road.[6] The problem was, in some ways, a simple one: there were more cars being driven more miles, as population and incomes increased after the war. In the United States especially, but also in reconstructed Europe by the 1950s, patterns of urban development and sprawl accompanying increasing home ownership led more people to drive more miles.[7] In the United States between 1950 and 1970, miles driven per year per capita increased at four times the rate of the population increase and twice the rate of increase of car ownership.[8] And the increase in car accidents seemed to outstrip even the phenomenal increase in car use. As public health statisticians and policymakers plotted the staggering increase in total miles driven, they were clearly worried about the declining safety of the roads.[9] Public health experts began to ask why accidents were occurring.[10] In the 1950s these researchers focused on crash avoidance by improving roads and cars and, most problematically, by educating people to be better drivers.[11] In 1958 the *Annals of the American Academy of Political and Social Science* published a book focused on the problem of road safety. Recognizing that the trend toward more cars was an unstoppable socioeconomic force, the editors called on the automotive industry to make safer cars.[12] Engineers were listening, and by 1958 both engineers and transportation policymakers were holding conferences to discuss what kinds of modifications to cars would be most beneficial. Skidding emerged as one of the more tractable problems, in part because researchers believed that skidding was not always the result of driver error. It was a problem which called out for a technological fix: better road surfaces, better tires, and most obviously better brakes and vehicle handling technologies. However, as is invariably true with technological fixes, the human element could not be eliminated for long, and the debate between making better drivers or making better cars began in the 1950s and remains a debate today.[13]

PHASE II: UNDERSTANDING SKIDDING

In their search for a technological fix, researchers had to define the problem of skidding. If skidding was to be prevented, researchers had to arrive at a common understanding of the causes of skidding. The initial work lay in empirical research that began before World War II. Engineers in the 1930s knew that skidding occurred when the friction between brake and wheel exceeded that between the tire and the road surface; consequently, the wheel locked up, causing the vehicle to slide. Generally speaking, skidding had three negative effects. First, on most surfaces, skidding increased the distance the vehicle required to stop.

Second, when a vehicle's wheels facilitated both steering and motive force, as they do on all automobiles but not, for example, on trains, skidding could cause the operator to lose directional control of the vehicle. Third, wheel lockup forced one area of a rubber tire to bear all of the friction of the stopping vehicle, causing the tire surface to heat up quickly and potentially blow out.[14]

In Britain skidding was perceived as a serious road hazard even before World War II. The newly formed Road Research Laboratory, a branch of the National Physical Laboratory, created a research program in the 1930s to investigate what happened when a car skidded.[15] Primarily, they wanted to determine whether having brakes on all four wheels would be beneficial, since it was not uncommon at the time for a car to be outfitted with a single drum brake on the rear axle and no front brakes at all.[16] The two main reasons to avoid front brakes were to keep the cost of the vehicle down and to prevent spin out, because many engineers and auto designers believed that front brakes increased the likelihood of spin out. In the 1930s engineers quickly determined that brakes on all four wheels would improve both directional control and stopping ability, but the car would spin predictably so that the locked wheels determined the direction of the skid. If the rear wheels locked, the car spun around backward; if the front wheels locked, the car simply continued in a straight line. Several articles blamed rear brakes for uncontrollable skidding. The easiest solution was to limit the rear brakes to a small percentage of the total braking force on the car.

This research was rudimentary compared to the work of engineers in the aviation industry who were also trying to understand skidding. In fact, the aviation industry was so far ahead of the automotive that the first commercially available antiskid devices appeared on airplanes in the United States right after World War II, nearly thirty years before their appearance on automobiles.[17] The development of skid preventing devices for planes began during the war. Skidding caused blowouts on aircraft during World War II, particularly carrier-based planes. These blowouts caused serious tire and rubber shortages for the U.S. Navy in the Pacific.[18] The supply problems were significant enough to interest the Navy in funding research to solve the blowout problem; Goodyear, Hydro-Aire, and Hayes Industries were the principal beneficiaries of these research contracts. One way to solve the problem of blowouts on planes was to keep the plane's wheels rotating; this was especially true for carrier-based planes, which were stopped by the carrier's tailhook.[19] This solution clearly did not apply to automobiles. For this reason, aircraft brake designers defined the problem they were trying to solve differently than would automotive engineers, whose concerns would be focused on handling and retaining control of the vehicle.

After the war, the initial impulse of automobile brake designers was to modify the aircraft brakes for application to cars. However, the much larger scale of a plane compared to a car proved to be an insurmountable problem. Aircraft devices could not be transformed or modified to work on cars; aircraft antiskid systems were no more the ancestors of ABS than the prewar patents.

After the failure of the direct transfer of technology from aviation, automotive engineers tried to understand the nature of skidding theoretically as well as experimentally. The most important locus for this research was Britain's Road Research Laboratory. Led by Director W. H. Glanville and Chief Engineer R. D. Lister in the late 1950s, RRL researchers focused on what the British referred to as "the interaction problem."[20] They hypothesized that skidding was caused by the interaction of the road, tire, automobile, and driver. Skidding accidents could be reduced by improvements in each of these areas. Glanville and his colleagues sought a combination of factors, including more effective tire treads, better road surfaces, better driver education (which consisted largely of a campaign to reduce drunk driving) and different designs for highway exchanges and road signs, as well as better braking systems. Although the reasons autos skidded could be broken down, skidding still remained a difficult problem to solve. In 1958 an international conference was held at the University of Virginia, where the program at the Road Research Lab was presented as a model for how to study skidding. This conference was the first international meeting of the skidding researchers and included thirty presenters and almost two hundred participants.

British engineers at the Road Research Lab emerged as the first experts in skidding research because the British government became involved in the problem of skidding long before any other. This happened partly because Britain was the only industrialized nation that kept nationalized statistics on skidding; in West Germany and the United States police reports and their resulting statistics were not standardized from state to state. When British statistical reports determined that skidding caused nearly a quarter of the road accidents, inventing ways to prevent single-car accidents, especially those attributable to skidding, became one of the most visible projects of the Road Research Lab. Furthermore, Britain in the 1950s was committed to government funding of applied scientific and engineering research, either directly in the case of the Road Research Lab or indirectly by funding industry research associations.[21]

Many of the ancestors of ABS were measuring devices rather than devices designed to prevent skidding. Institutions such as the Road Research Lab were focused on instrumental problems, developing better testing instruments, not on making a commercial antilock braking system. This is one reason why the institutional diversity of the ABS knowledge community was one of its strengths. The development of antiskid devices often relied on new technologies coming from the testing of road surfaces and tire treads. Knowledge generated by the invention of measuring devices played a critical role in solving the considerable difficulty of making real-time measurements, which is the central function of an ABS. At its core, ABS acts as a complex measuring device, constantly measuring and comparing the instantaneous rates of deceleration in two or four wheels of a vehicle. So the history of ABS begins not only with people trying to develop skid prevention, but also people trying to make the measurements that would explain the problems of skidding. It turns out that skid measuring instrumentation, even if not as a component of a braking system, played a much greater role in ABS development than did designs for antilock braking systems of the 1930s.

PHASE III: SPECIALIZATION AND COMMERCIALIZATION

By the late 1950s RRL chief engineer Lister had broken out of the intractable interaction paradigm and was once again experimenting with rigging an aircraft antiskid unit to a car. He did not expect this application to be viable; it was an experiment to help him understand what functional qualities an automotive antiskid device needed to have. Lister retrofitted a Dunlop Maxaret aircraft antiskid unit to a 1950 Morris 6 automobile. This required several variations to the car; for example, the Maxaret worked only with disc brakes, which were standard in the aircraft industry but uncommon on production automobiles, which used drum brakes. Modifications were made to both the auto and the Maxaret unit, and ultimately Lister and his collaborators at Dunlop used this knowledge to design a functional, producible device for passenger cars. However, the automotive Maxaret antiskid device cost over £2,000 in 1965, the cost of about four VW Beetles. The Maxaret was available as an option on the £7,000 Rolls-Royce Silver Shadow and the innovative Jensen FF, as well as an assortment of Jaguars. In addition to its prohibitive cost, the Maxaret required four-wheel disc brakes, a particular suspension, and a different hydraulic system than most production automobiles had.

The Maxaret, developed by Lister in conjunction with Dunlop's chief brake engineer, J. W. Kinchin, was one of the earliest legitimate ancestors of present-day antilock systems.[22] However, it is easy to overstate its impact. Statistical research showed that in order to reduce skidding accidents, skid prevention had to apply to a broader market than Rolls-Royces and Jaguars. One could safely say that installing skid prevention on every Jensen FF in the United Kingdom would make no statistical dent in skidding accidents, since only 330 of the hand-built cars were produced between 1965 and 1970. Also, the Maxaret never worked flawlessly; it could be inaccurate in detecting wheel speed and often reacted too slowly.[23] The Maxaret was most successful as a paradigm for how to do antilock R&D, and Lister published several papers on its development. But the Maxaret did not solve the problem of skidding.

Meanwhile in West Germany, at Teldix GmbH, a small aviation electronic company in Heidelberg, a research team was working on producing small, high-speed, high-pressure hydraulic valves for the gyroscopic and guidance system of the F104 Starfighter military jet. Starting in 1964, Heinz Leiber, a Teldix engineer, sought to apply this valve to an automotive system as a solution to the problems of the Maxaret brake. Since antilock braking systems pulsed the brake caliper very quickly, a small, high-speed valve was an essential component. The valves in the Maxaret could pulse six times per second; Teldix's valve could pulse sixty times per second. In 1969 Leiber secured a contract with nearby Daimler-Benz AG to develop an antilock system based on his valve.[24] It was Teldix's well-publicized relationship with Daimler that sent executives from Bosch (like Daimler, located in Stuttgart) up the road to Heidelberg to buy a 50 percent interest in Teldix in 1973.

While Teldix engineers did the work that Bosch would be known for, a company in Frankfurt was performing important research and development that led to another successful market introduction in 1978. Somewhat surprisingly, with only a few exceptions traditional brake producers avoided work on ABS, leaving its R&D to automotive electrical firms, such as Bosch, which had developed inexpensive, production electronic fuel injection in the 1960s. Alfred Teves, GmbH, was the most important brake producer to enter the ABS race. Hans Strien, a senior engineer at Teves in the 1960s, had been instrumental in developing brake pressure distribution systems in the 1950s.[25] Strien expanded his research into antiskid systems in the early 1960s. He knew that the defining problem of ABS would be its cost, so he and his colleagues at Teves worked on the production of smaller and smaller hydraulic valves and fewer components in order to reduce the system's potential cost.[26]

Strien regularly published in English and French engineering society journals as well as through FISITA, the International Congress of Automotive Engineers. In this way he helped to reinforce the international character of ABS work, established himself as an expert in the international field, and was privy to much new knowledge coming out of smaller companies.

In addition to Teves, other companies were developing competing systems without the benefit of the Teldix valve. In France in 1973 the Bendix subsidiary DBA was readying an ABS called Autostable. It was to appear on a Citroën, but the oil crisis halted production of the car, and neither that car nor the Autostable ever made it to market.[27] A French, Italian, and British collaboration at the Fiat Centre for Electronic Research introduced a successful ABS for commercial vehicles, but their system for passenger cars was legally challenged by an American company as an international patent violation; the legal battle prevented the system from being introduced.[28] Neither of these two important failures were ever tested in the market. Production costs for all antiskid systems were initially very high, and these aborted attempts never moved into mass production. Nevertheless, their importance to the engineering community is expressed through the number of papers contributed to the field. Of the over one thousand papers written on ABS in the 1960s and 1970s, almost all presented systems that were ultimately unsuccessful on the market for both technical and economic reasons.

The unsuccessful system of the greatest importance to the field in the late 1960s was a product of the research program of the VTI, the Swedish traffic institute, Statens-väg och trafikinstitut. The VTI was modeled after the Road Research Laboratory in Britain; it encouraged collaborative research between Swedish companies by funding a large-scale project at a government-owned facility in Linkoping. Unlike the Road Research Lab, the goal of the VTI was to produce devices, which it patented in Sweden and abroad. For obvious reasons, Swedish engineers were most interested in tackling the problem of skidding on ice. This constituted the extreme case, because friction between an icy road and the tire of a vehicle is minimal. Gösta Kullberg, Olle Nordström, and Göran Palmkvist undertook a fifteen-year study to produce optical, as opposed to mechanical, sensors. Their purpose was to "build up technical know-how independent of industry concerns and offer the results at the disposal of industry."[29] However, Sweden's role is far more important than the work done at the VTI. Starting in 1973 the world's largest brake proving grounds were developed in Sweden, supported by the VTI. As legend has it, German automotive engineers were visiting the town of Arjeplog, an old silver mining town in Swedish Lapland, just below the Arctic Circle. They saw a winter runway on part of Lake Hornavan and asked whether they might use the cleared section for some brake testing; word of the ideal conditions for ice tracks quickly spread among automobile engineers.[30] By 2004 the lake was the site for more than 1,800 kilometers of ice track, and the testing industry provided two thousand wintertime jobs, yielding the lowest unemployment rate in all of Sweden.[31]

So far the development of ABS as I have told it centers on the development of knowledge communities and artifacts. However, the growth of Arjeplog shows that institutions (and the facilities that institutions provided) also mattered. While corporations obviously had a considerable stake in producing ABS—it was a product that promised profits—public institutions, particularly government agencies like the Road Research Lab and the VTI, proved to be important sites for both the production and the dissemination of knowledge. A number of universities also took a particular interest in skidding as it could be applied to the embryonic science of vehicle dynamics, which was growing out of aerospace engineering. In addition, several universities all over the world set up research institutes for traffic safety; the largest were at Braunschweig, Munich, Berlin, Delft, Uppsala, Belgrade, the Royal College of Aeronautics in Britain, the University of Michigan, Cornell University, and Penn State University. The universities contributed important theoretical tools, but more important, they provided an environment for discussing research. The university atmosphere increased the engineers' sense of professional responsibility in sharing important information that was loosely construed by the participating engineers as nonproprietary, that is, general information that was either not patentable or was already patent-protected.[32]

By the mid-1970s most companies, as well as the Swedish government, had entered into a more proprietary phase of work and less information was being shared. The number of publications dropped sharply each year after 1973 and research articles appeared in more specialized venues, such as the annual conferences devoted to ABS held by the Institution of Mechanical Engineers in London. Two devices became commercially available in the 1978 model year: Bosch's and Teves's, appearing on Mercedes-Benz and BMW cars, respectively. The period of the mid-1970s was marked by increased attention to mass-production schemes. For example, Teldix's research was entirely given over to Bosch in 1975, in large part because Bosch had a solid track record in mass production, while Teldix had produced only small batches of avionics devices. Other companies were carving out a niche at the bottom end of the cost scale; this was largely the strategy of Lucas, an auto electrics company in Birmingham, England. Lucas resisted the trend toward digital electronic systems to develop a much less expensive mechanical system, the SCS. In contrast to the £1,000 Maxaret system of 1965, the first generation of Lucas's SCS, their most expensive system, cost less than £400 a decade later.[33] Although Lucas's lack of success with the SCS was part of the tragic end of British automobile engineering and manufacturing in the 1970s, its strategy showed that room existed within the ABS community for different approaches. Paul Oppenheimer, one of the inventors of the SCS, became the architect of regulatory schemes for ABS in the European Economic Community and the UN's Economic Commission for Europe.

Although the knowledge community approached projects with less single-minded intensity in the 1970s than it had in the 1960s, it also had greater maturity and diversity. The marketing of ABS after 1978 helped to sort out the successful systems from the failures and brought back a coherent focus, seeking increased reliability, decreased cost, and exclusively digital electronic systems.

MOTIVATIONS FOR ABS DEVELOPMENT

Explaining the development of ABS through its genealogies provides some insight into the motives of engineers, which remain an open and complex question in the case of ABS, especially given the large number of unsuccessful systems that engineers produced. While there is an obvious financial motive for corporations that decide to pursue a new technology, engineers' motives are less clear and one-dimensional. A corporation such as Bosch might be convinced that a long-term investment in research and development will pay off, but engineers might look at the same situation with less patience. The motives of engineers were supported by the knowledge community, which served to highlight certain problems for special attention. These problems, which are detailed in the following chapters, acted as puzzles and attracted the best minds of the knowledge community. The overarching problem was to create a system that both shortened stopping distance and allowed drivers to retain directional control. But smaller, more focused problems were also highly valued, and they constituted the problems around which the community congealed. These intractable puzzles served as the attractors for the community as well as being defined by the community; the circularity inherent in this conception is not a logical mistake, but rather the reason why the social and epistemological dimensions of the community were inseparable. The problems also served a social function in that solving them, even when such solutions did not constitute a completely new and functional ABS, promised rewards and recognition by fellow members of the knowledge community.

This social structure of the engineering community, which highlighted particular research topics and defined practices, resembles the paradigm that Thomas Kuhn discusses in his work on the social dimensions of scientific research. Kuhn was looking for a way to express what members of a scientific community share, their way of thinking. He suggested that the term *paradigm* might be fruitfully replaced with the phrase *disciplinary matrix*, which includes symbolic generalizations, models, and exemplars.[34] In *Origins of the Turbojet Revolution* Edward

Constant applies Kuhn's work to technology and engineering communities. Constant posits that these communities share "traditions of practice"; these include "relevant scientific theories, engineering design formulae, accepted procedures and methods, specialized instruments, and often elements of ideological rationale."[35] Traditions of practice contain both a knowledge dimension and a sociological dimension, which includes an informal organizational structure with behavioral norms. Constant writes that the turbojet community shared a tradition of practice, as did the propeller-design community. The scientific revolution that Kuhn described occurs when the jet tradition becomes dominant and the propeller tradition fades into a secondary role.[36] This constitutes a shuffling of the puzzle pieces and a revaluing of new puzzles posed by jet engine production. As Constant's work has shown, the development of the jet is an appropriate case study for applying a Kuhnian model to technology; no other attempts have worked as well.

Giovanni Dosi's concept of technological paradigms is very similar to Constant's. Dosi's most important contribution is the notion that a paradigm defines its own concept of progress, based on a system of trade-offs between specific technological and economic goals.[37] Such definitions of progress motivate engineers to solve obstacles. Through this negotiated progress, and corresponding improvements in devices and processes, a technological paradigm brings with it a specific technological trajectory. The function of the trajectory is prescriptive, in that the trajectory will exclude certain solution sets. The trajectory not only makes some alternatives more attractive in terms of its internally defined notion of progress, but also eliminates certain potential resources from the knowledge pool. The way the goals or questions are posed delimits a solution set.[38]

Considering the way knowledge communities create motivations for certain research problems is an important piece of what Michael S. Mahoney refers to as a community's agenda. Focusing on motivations also pushes economic issues into the mix, which is important to understanding research and development in the private sector. The engineers who spent decades on ABS were motivated by an economic reward structure, but that was often overshadowed by more personal and psychological motivations that the knowledge community attended to and reinforced by offering problem solvers awards, expert status, and publications.

three

The British Road Research Laboratory

Constructing the Questions

In September 1958 over two hundred researchers met in New-
comb Hall at the University of Virginia for five days to discuss
an exploding international public health and safety problem:
automobile crashes caused by drivers who lost control of their
vehicles while skidding.[1] The escalating number of passenger
automobiles, as well as the increasing volume of commercial
traffic, underlay the frequency and severity of skidding-related
accidents, and there was no relief in sight. Automotive, civil,
and mechanical engineers, urban planners, and highway department officials brought different proposals and preliminary
test results to share. Road surfaces, highway layout, brake shoe
materials, tire tread and composition, braking system design,
and driver education all presented possible avenues for reducing traffic accidents.

The first International Skid Prevention Conference was organized by Tilton E. Shelburne of the Virginia Council of Highway Investigation and Research and Dr. W. H. Glanville of
the Road Research Laboratory in Slough, England. The purpose was to provide a forum for the exchange of information
among different countries' representatives from various specialized fields, all of whom were working on various aspects and
phases of the skidding problem. Shelburne, a civil engineer
interested in road surface design, emphasized the need to share
information about the measurement of pavement slipperiness,
as several different methods had been developed since World
War II. When he learned that similar but incommensurable

metrics were developing in other areas of road research, he called for "a survey of what is now known and a clarification of what we need to know" in 1957.[2] Glanville brought a different but related set of questions to the conference, as his research focused on making the automobile itself less likely to skid, particularly by improving the design of tire treads and composition. Glanville and Shelburne knew each other through their similar administrative positions with the Virginia Council and the RRL, but their research specialties put them into two different knowledge communities. Their administrative connection led to discussions about the common questions researchers in their respective fields were posing; in both fields questions were being raised about how to measure the problem of skidding. In fact, though asked by different communities, their research inquiries did overlap; tire research was inextricably connected to pavement engineering because most research was concerned with finding the coefficient of friction between the road and the tire. Glanville and Shelburne, although they had specialty knowledge, both provided a multidimensional approach to the problem of skidding. They sought to carry this approach to the participants of the conference.

Shelburne and Glanville assembled a committee of twenty individuals from government agencies in Britain and the United States, including the Highway Research Board, the Bureau of Public Roads, and the National Safety Council. Industries were well represented on the steering committees, often by directors of industry collectives, including those representing the tire, asphalt, Portland cement, crushed stone, automobile, and truck-trailer industries. This committee organized the conference into five subcommittees: vehicular and brake design, accident analysis and the human element, tire design and composition, road surface properties, and laboratory and field methods for measuring road surface friction. Shelburne, as an official of an American highway agency, oversaw the programs related to road surfaces, while Glanville concentrated on the first three subcommittees, studying brake and vehicle design, accident data, and tire design. Glanville was considered an outstanding expert in these broad areas because his agency, the British RRL, was the world leader in research on these three areas.

The conference impressed those attending it with the complexity and potential power of the RRL's method of investigating skidding. The RRL's program was set up exactly like the conference, with subcommittees of researchers working on each of the five categories. The concept of skidding as a problem of "the interaction between road, tyre, vehicle and driver" was accepted as a viable path to finding solutions through better driver training, better road signage, higher coefficients of friction between road and tire, and better braking systems.[3] Comments at the conference keyed on the RRL's ability to study problems at the government's expense, with a five-year research timetable, and to look at far more aspects of the skidding problem than any other agency had access to.[4] Between 1946 and 1960 the RRL spent more time, money, and effort on the skidding problem than any other research agency or private company. Consequently, RRL members occupied positions of prominence at the Skid Prevention Conference. Several participants commended the RRL on its role and the quality of its work in the field of skidding: "[The RRL's] permanent, suitably qualified staff, [is] so organized, financed, and positioned that it is free to study any road or highway matter including safety—and apparently to devote reasonably adequate time and consideration to selected researches. This English organization is clearly effective in cooperating with private industry research and development, and in getting highway authorities and others to apply research findings with outstanding results."[5] Yet by 1963 the RRL had virtually vanished as a source of information on every element of skidding research but the question of road surface. In this chapter I investigate the factors that explain the RRL's interest, then its success, then its disappearance from the scene of skidding research. No other agency framed the problem of skidding as one of the interrelated factors of road, vehicle, and driver.

THE RRL: EXAMINING THE INTERACTION BETWEEN ROAD, TIRE, VEHICLE, AND DRIVER

On behalf of the British Ministry of Transport, the National Physical Laboratory created a research program in 1911 to investigate the problem of skidding in automobiles. After a hiatus during World War I, the program was restarted by the Department of Scientific and Industrial Research (DSIR) and in 1933 was given a facility at Harmondsworth, Middlesex.[6] The DSIR ran research programs for various government agencies, among them an investigation into skidding for the Ministry of Transport. This program was called the Road Research Programme and used model cars to investigate the relationship between the motion of a vehicle and which of its wheels were locked.[7] These simplified, minia-ture vehicles were used to keep down cost and because the facility at Harmondsworth did not have grounds sufficiently large enough to skid full-size cars at high speeds. By building a small, rigid chassis with four wheels and four brakes of the two-shoe, internal-expanding variety, the engineers were able to test empirically the dynamics of the vehicle on which each wheel could be independently locked.

This miniature vehicle was towed four feet, then let go to travel under its own momentum. Brakes were applied and the vehicle came to a measured stop. In this way the engineers discovered only what had already been known: the shortest braking distance with the least devia-tion occurred when the cart stopped with all four wheels locked. But would this model be satisfactory for full-scale vehicles? Were there scale effects in translating the behavior of the miniature to a full-size car? Were low-speed effects proportional to phenomena generated at actual driving speeds? What happened when the chassis was not rigid, when the vehicle had a suspension and a steering system? What other vari-ables did the miniature ignore, and what role did they play in determin-ing the behavior of an actual vehicle? These were the questions that Road Research engineers asked when work resumed after World War II, but they were questions answerable only by comparison to tests done on real cars at driving speeds.

Work resumed in 1946, and a new Road Research Laboratory was set up in Slough, Berkshire, to continue the work of the Road Research Programme. The RRL retained its prestigious affiliations with the Minis-try of Transport, the National Physical Laboratory, and the DSIR, and it also gained a facility on which to run full-scale, full-speed experiments. This made the RRL the most important and best endowed road research facility in the world. The RRL's main mission was to act as a testing facility for road design.[8] In their investigations of the problems occur-ring on roads in Great Britain, researchers found that road design was often the entry into a set of what were termed "interaction problems" and they became interested in figuring out what roles various inter-related factors—drivers, roads, and vehicles—played in causing acci-dents. Often their finding was that accidents were caused by the inter-action of these factors, thus making them particularly difficult to remedy since the elimination of a single factor might not completely alleviate a given danger. This focus on problems of interaction led researchers at the RRL to be concerned with questions of whether road dangers were particular—that is, confined to a single location and amenable to a local solution—or general, owing to such factors as the greater volume of traffic and its higher speeds. Most of their effort was focused on the latter. For most of the 1950s the RRL was the only organization in the world with the access and the voice to make improvements on a national scale; as a result, by 1955 the RRL was a model for other national investigation bureaus.

WHY SKIDDING?

The RRL's interest in skid prevention began with statistical findings indicating that skidding accidents had increased between 1951 and 1957. In 1957 alone, 16,000 accidents were caused by skidding on wet surfaces, representing 27 percent of accidents under wet conditions. Skidding accounted for a much higher percentage of accidents on icy or snowy roads: 82 percent, but because of the relatively mild climate of Great Britain, these 3,000 accidents constituted a much smaller percentage of the total number of vehicle accidents. Skidding was a relatively minor problem on dry pavement, contributing to only 8 percent of such accidents but accounting for a total of 15,000 accidents. In all, out of a total of 216,000 injury accidents in 1957, over 30,000 involved skidding.[9] Since neither Germany nor the United States kept national accident statistics at this time, Britain's findings were taken as representative for industrialized nations.[10] Nations of a more northern climate, such as Sweden, would face much larger percentages of snow- and ice-related skidding.[11]

Few of the RRL's reports tried to explain the reason for the increase in skidding-related accidents. Although traffic volume was increasing, the rate of skidding-related accidents outstripped the increase in vehicle miles. Generally speculation regarding this growth pointed to the more powerful engines installed in cars, which otherwise remained very much the same as their prewar ancestors. Speed was well-established as a disproportionally important factor in skidding on wet roads because of a phenomenon known as hydroplaning. Hydroplaning occurs when the tire is unable to channel water out of the contact plane between the road and the tire. As a result the tires literally skim along the water's surface, leading to a very low coefficient of friction. The higher the speed, the more likely a vehicle is to hydroplane.[12] Furthermore, because on wet surfaces the coefficient of friction increases linearly with speed while the kinetic energy of the vehicle varies with the square of the speed, the distance that a vehicle requires to stop is much greater on a wet road.[13] As a result, higher average speeds would partially explain a greater number of skidding-related accidents, at least on wet surfaces. O. K. Normann, who looked at figures in the United States, found the average speed to be 16 miles per hour higher in 1957 than in 1941, causing a 78 percent increase in kinetic energy and a 78 percent increase in braking distances.[14] The RRL had similar statistics that prompted particular concern over skidding rates on wet roads. Yet reducing driving speeds in the face of engine technology that made increased speeds easier to achieve and sustain posed a difficult problem for government law enforcement and highway agencies.

The RRL's attention to skidding was also prompted by the sense that skidding was the one significant cause of accidents that held promise for being more or less correctable. The most statistically significant cause of accidents everywhere was driver error. While there was some debate as to whether the rate of driver errors could be decreased through various technologies, these errors were usually considered uncorrectable by engineering. In addition, minor driver offenses, such as driving too fast, tailgating, failing to yield, and careless passing, often led to accidents when combined with skidding on occasions where the careless behavior might otherwise have been uneventful. If skidding could be prevented it would impact numerous other causal factors and perhaps play a disproportionally important role in making roads safer. Furthermore, while changing driver behavior looked impossible, skidding seemed potentially amenable to improvement through engineering by making better road and tire designs as well as better vehicles and braking systems. In the 1950s public trust in the technological fix was riding high. Fixes aimed at braking systems got a particular boost from the well-publicized antiskid devices on aircraft that had significantly reduced the problem of planes skidding less than a decade earlier, in the late 1940s.

Despite the early promise of braking technologies as potential solutions to skidding problems, it was not the only avenue of development the RRL pursued. In fact, there was no perceptible preference for any one area of solutions over any other at the outset of the post–World War II research program in skidding. Because Glanville officially attributed skidding to a conglomeration of factors referred to as the "interaction between road, tyre, vehicle and driver," all these areas warranted equal attention by the various research teams of the RRL. However, the notion of an interaction problem also caused difficulties, because although it captured the interdependency of the phenomena, it also posed an intractable question: Where did one begin to unravel this puzzle? What was the first step, the most important dimension, the prime mover? An interaction problem required a multidisciplinary organization like the RRL, since no company in the world dealt with as many causal factors. Even a large corporation, such as General Motors or Dunlop Tire, was not in the business of road design. So corporate research necessarily defined the problem of skidding in different ways to make it tractable, given each company's resources and interest.

Despite the RRL's focus on the multidimensionality of skidding, the agency still broke its interaction problem down into subproblems to construct research agendas. This project of parsing a grand conceptual problem into concrete, directable parts is what Walter Vincenti refers to when he discusses design as a multilevel, hierarchical process, in which the problem must be parsed from its most vague and unstructured commercial requirements into major components and then into highly specific details.[15] The smaller problems can be solved in relative isolation from the rest of the project. Naturally, reintegrating the components can pose major problems, but without an analytical approach to the system, research agendas are difficult to determine. Examples of these subproblems at RRL were investigations into applying various types of grit to roundabouts to increase traction, road signs that could advise drivers of changing climatic conditions, and attempts to install an aircraft antiskid device on an automobile for testing. At the RRL the interaction problem was broken into at least six units. The highest profile unit performed experiments to determine the coefficients of friction between various tire and road surfaces. Another unit investigated the process of marking roads for the purpose of better informing drivers. A third unit looked into the condition of cars currently in service. A fourth unit looked into improving the performance of braking systems either by add-on devices or general improvements to the system. Other units investigated driver psychology with the hopes of improving driver training, vehicle dynamics regarding the interaction between braking and cornering, snow and ice removal systems, and the physics of friction.[16] Glanville's great achievement was in keeping these semidiscrete units apprised of each other's progress and allowing some participants to work on more than one project. In doing so, Glanville followed a communal model of R&D espoused by the RRL's parent agency, the DSIR.[17]

The degree to which RRL researchers thought of skidding as an interaction problem is seen in the fact that no one project area claimed that if its recommendations were accepted the problem would disappear. Even the most confident propositions were accompanied by caveats "that only the driver can truly prevent skidding" or "that no road surface will solve the skidding problem under unfavorable climatic conditions."[18] James Stannard Baker of the Northwestern University Traffic Institute reported on several case studies to highlight truly bizarre sources of skidding accidents. These situations could neither be anticipated nor corrected by either the engineer or the driver. The oddest case involved a swarm of black grasshoppers which took up residence on a Utah highway for several days one summer, perfectly camouflaged against the black pavement. Several people, driving within the speed limit, hit the swarm and either spun off the road or hit other vehicles. No driver could anticipate this situation, in which a very dry road suddenly became soaked with rather slippery grasshopper innards.[19]

TIRE TREADS AND ROAD SURFACES

Since the friction between a tire and the road depends on the surface of both, the road engineers and tire engineers at the RRL worked as closely together as any two groups. While road specialists and tire engineers had different industry audiences and different objects of inquiry, increasing the friction between road and tire required a great deal of collaborative work. Changes in the standards for road surfaces had to take into account the most common types of tires being used by motorists; conversely, improvements in tire tread were made with respect to commonly used road materials. It did no good to design either road surfaces or tire treads independently, since the key variable was the friction between the two surfaces, with a considerable concern for the wear characteristics of each. Cyril George Giles, the RRL's chief researcher of road and tire properties, wrote in 1953:

It will be seen that in spite of the progress that has been made it is still the problem of skidding on wet roads which is of greatest importance, but the emphasis is changing. The old smooth-looking surfaces which used to prove dangerously slippery in wet weather are fast disappearing and today the main concern is more and more with those coarse-textured surfaces with a rugged "non-skid" appearance which have a relatively low resistance to skidding because the stones in the surface have been polished. On the older smoother-looking surfaces, tyre tread patterns, by providing drainage can play an important role in resisting skidding and the main need today is for tyre characteristics which will be as effective to a similar degree on modern, rough, coarse textured surfaces. A solution to this problem, if it could be found would be a major step forward in the direction of safety.[20]

In practical terms, the RRL's recommendations for surface improvements on roads required knowledge of the past (as cars could have tires up to about ten years old), the present, and the future direction of tire technology. This drew on the RRL's relationship with the tire industry, as well as its own surveys about the condition of vehicles in use. Even when improvements were made in the area of tire-road friction, the effects were limited in scale. For example, a 33 percent improvement of the coefficient of friction decreases the stopping distance of a 30 mph car only 25 feet. As a result Giles's group looked for more specific "skid zones," particular sections of road where skidding had proven to be a problem over time. Testing novel road surfaces at limited locations with known problems was a cost-effective and tractable way to test the efficacy of potential solutions before the solutions were committed to wider usage. Giles also hoped that by attacking known problem spots, his program would have a large effect in reducing skidding accidents.[21]

"THE DRIVER MUST BE ASSISTED BUT WILL NOT TAKE KINDLY TO BEING REPLACED"

The group that faced the greatest difficulty in separating their task from the other areas was the team of researchers focusing on the problems of driver judgment and error.[22] This group also faced particularly heated political battles for acceptance of their recommendations. Drivers approved of reducing the number of accidents, but they did not want the techniques and "feel" of driving to be significantly changed. Any suggestions that the driver's instincts be overridden by technology received a particularly acrimonious response from members of the Institution of Mechanical Engineers.

Despite this opposition, the achievements of driver psychology researchers probably made the largest difference in improving the safety record per vehicle miles, particularly those targeting the problem of intoxicated driving. In the 1950s human engineering specialists backed changes in the hands-on training of drivers, road signs, speed limits, and legislation against driving while intoxicated. In addition, many of these engineers worked to improve the ergonomics of the vehicle to lower driver reaction times and increase feedback from the vehicle to the driver, especially in steering systems.[23] Despite the success of these efforts, some members of the Institute of Mechanical Engineers resisted change. Many respected engineers fought against road signs in particular, believing that signs would be ineffective and might even discredit the original designers of roads. The best example of this is a comment made by S. B. Bailey in a discussion recorded by the Institution of Mechanical Engineers. Bailey recounted a story about the effectiveness of both speed limits and warning signs: "He had seen an articulated tank transporter carrying a load of 100 tons and displaying a large sign reading 'DANGER. LEFT-HAND DRIVE. NO SIGNALS. MAXIMUM SPEED 12 M.P.H.' It was proceeding at a steady 40 M.P.H. [Bailey] suggested that it would have been better to take the notice off the back and put one on the front reading, 'DANGER. NO SENSE. NO BRAKES. CANNOT STOP!'"[24] Similar stories arguing that signs had little benefit and potential unintended consequences were recounted throughout the pages of the Institution's proceedings. Despite this sentiment, many of the recommendations for driver training, legislation, and road marking had become law by 1970.

WHAT'S A BRAKE TO DO? THE RRL AND ANTISKID DEVICES

One of the fortunate consequences of the DSIR's highly communicative style of work programs was that Glanville allowed researchers to slide between projects. Research into the improvement of braking systems was a particular beneficiary of this flexibility. R. D. Lister, employed by the RRL in 1946, first worked on surveying the condition of autos currently on the road.[25] His study provided many challenges in developing methods of testing, which would play a large role in his later research on antiskid devices. The first and perhaps most obvious problem was how to measure the stopping ability of a vehicle. Over about a decade,

Lister and others at the RRL came up with seven different methods for measuring braking distance and calculating deceleration. They gathered enough data on each method to calculate standard deviation figures for each method. The most accurate instrument became the hallmark of the RRL braking investigation program, the fifth wheel apparatus.

The previous standard of accuracy in measuring stopping distance was direct measurement. In this method, a switch attached to the brake pedal completed a circuit when depressed. That circuit fired a charge, which left a chalk mark on the pavement within four inches of the spot where the brake pedal had been engaged. The distance between that mark and the final, stopped position of the vehicle was then measured. Because it was highly accurate this method was the standard against which all other methods were compared. However, it required that the vehicle's brake system be adapted to include the switch, so it was not practical for cars in everyday use. The fifth wheel method, on the other hand, used a trailer which could be attached to any vehicle in a couple of minutes. The vehicle was driven down a track and a bicycle wheel on the trailer measured the distance between the initial application of the brake pedal and the point at which the vehicle actually stopped. The fifth wheel apparatus was triggered by a magnet embedded in the road a few feet after a sign told the driver to begin braking. The wheel then counted the number of revolutions it completed until the car came to a complete stop. This method yielded less than 1.1 percent deviation from the direct measurement method and was a practical solution for testing vehicles randomly selected off the street. Other methods were devised to measure the braking force at each wheel, though theoretically the amount of force could be derived from stopping distance.

Lister's findings in this program told researchers a great deal about the quality of braking systems in cars currently on the road and led to further research into improving braking systems. In general, the RRL found that in a significant percentage of vehicles tested, the condition of braking systems was horrible. In a random check in 1950, the average stopping distance at 30 mph was 70 feet. The shortest stopping distance, achieved with all four wheels locked, was 43 feet.[26] Ten percent of cars could not stop within 100 feet and 1 percent could not stop within 200 feet. The average maximum deceleration achievable was 0.57g. Amazingly, over 25 percent of all vehicles had at least one wheel with no braking force at all. These results were especially upsetting considering that well over half of the vehicles tested were postwar models. Lister took these results to the human engineering group. But like most problems the RRL attended to, this problem had multiple causes: Were the cars in such poor condition because of their design, because of poor or insufficient maintenance, or because the drivers chose not to have repairs made? The answer lay with manufacturing quality control and upkeep of the vehicles; driver education could address the latter. But these poor results distressed Lister, as can be seen in his subsequent search for devices that could reduce skidding without the knowledge of and without being activated by the driver.

ENTER DUNLOP

In 1953 Lister obtained an aircraft antiskid unit from Dunlop called the Maxaret system. The Maxaret had been developed by a team headed by J. W. Kinchin that appeared on the aviation market in 1952. It was accompanied by very similar devices from Hydro-Aire, Lockheed, Westinghouse, and Goodyear, all of which hit the market between 1949 and 1952.[27] The aircraft antiskid systems were designed to prevent wheel lockup for both economic and safety reasons. On a landing plane a locked wheel could cause a tire to develop a flat spot after only one second of skidding and could blow out before three seconds passed. Blowouts were extremely dangerous as they often sent the plane careening off the runway. The weight of the plane and the pilot's position made skidding impossible to sense, eliminating the option of training pilots to modify their actions in a skid.[28] In addition, the tires on aircraft were (and still are) very expensive, and airline companies usually planned on retreading them a dozen times or more. A blowout destroyed the tire, so frequent blowouts were very costly. Other devices were invented that electrically prerotated the wheels of a jet so that the lower coefficient of static friction would not occur on touchdown.[29]

Dunlop's Maxaret unit consisted of a small rubber-tired wheel positioned against the inside rim of the vehicle's wheel. Inside the wheel were a drum, flywheel, and drive spring. The spring held the drum in contact with the flywheel. When deceleration exceeded a certain preset point the spring retracted and the drum stopped rotating. The flywheel, which had 60 degrees of free rotation, continued to rotate relative to the drum. A valve was triggered by this action to close the pressure supply line to the brake-actuating cylinder. Pressure was then exhausted from the actuating cylinder. The brake was released until the wheel regained its angular velocity, at which point the spring reset itself and the cycle started over. This system could generate 6.5 to 7 cycles of braking and releasing per second.[30]

Lister installed the Maxaret on a 1950 Morris 6 automobile equipped with hydraulically actuated, self-energizing drum brakes on front and rear. This involved significantly adapting the device, because the Maxaret was designed to work with disc brakes. Since each wheel had a unit, braking distribution problems were eliminated.[31] The system simply detected unacceptable deceleration and modulated pressure at that wheel; the system had no capacity to compare the deceleration or braking pressure of the front and rear. Both treaded and smooth tires were tested. Lister also modified the hydraulic assist system of the aircraft Maxaret to allow hydraulic fluid to continue to pass through during modulations, so the experimental setup was less powerful, as was appropriate for a much smaller vehicle. The threshold, or point at which the Maxaret engaged, was set at 0.65g, a much higher deceleration rate than the unit used on aircraft.

The tests were very successful in retaining directional stability, but not in shortening braking distance. On all but a few surfaces, braking distances were actually lengthened by the Maxaret because it could not keep the braking efficiency high enough. Braking efficiency refers to the amount of braking the system is actually doing. For example, if one steps on the brake pedal and presses it as far to the floor as it will go, the automobile's brakes should fully engage. In this situation, the braking efficiency will be 100 percent and the wheels will lock. However, with the wheels locked the driver will completely lose directional control and a slightly longer stopping distance will result than if the brakes were engaged to just less than the point at which the wheels lock up; this is because the coefficient of static friction is less than the coefficient of rolling friction. But the Maxaret initially led to a different problem: it did not engage the brakes enough. Thus braking efficiency was too low, as though the brake pedal were only partially engaged. Under these conditions, the car's wheels locked only for an instant, so directional stability was retained but the distance it took the car to stop was lengthened. Still, it was promising that the few conditions under which the braking distances were actually shorter were those encountered when the vehicle was stopped from its highest speeds or on wet pavement. Nevertheless, longer braking distances, even at low speeds, were not acceptable. Lister's experiment had pinpointed areas needing further work.

There were additional problems with Lister's initial attempts to retrofit an aircraft Maxaret to an automobile. In general, the system did not respond quickly enough or to fine enough gradations of wheel slip. These difficulties were directly traceable to the way the device performed on aircraft; on an airplane the device did not need to respond to mild wheel slip. In addition, the Maxaret was too large for automobiles and made the steering of the vehicle sluggish. If a device of the Maxaret's type were to succeed at preventing wheel slip in autos, it would have to be designed specifically for the automotive market, which opened a debate about what kind of automotive braking system the device should be designed to work with.

There have always been a great variety of automobile braking systems, encompassing different actuating systems (pneumatic and hydraulic), different types of brakes (principally disc and drum), and different numbers of brakes (either on each wheel or on each axle). Consequently, antiskid devices might have to accommodate multiple types of systems, and this in itself posed problems. European car manufacturers came early to prefer the disc brake, and many British engineers were convinced that the disc brake would improve the questionable braking performance exhibited in Lister's survey of in-use vehicles. The disc brake was common on European luxury and sports cars from the 1950s, but less expensive cars continue to use drum brakes, at least on the rear, to the present day. If antiskid devices worked only with disc brakes, then only very expensive models, such as Jaguars and Rolls-Royces, would have them, but in order to eliminate or even reduce the hazard caused by skidding the devices would have to become much more common. Engineers also determined early on that given the diversity of braking systems in use, antiskid devices could not be generic; that is, each antiskid system would be designed to work with a particular braking system. In the case of Dunlop and the RRL's collaborative research on the Maxaret, the device was predicated on the use of the four-wheel disc brake system, a rare and expensive system, but one which was advocated by Jaguar in the 1950s.[32] Lister knew that making the device work in the first place would be tough enough, but making this kind of system affordable to the masses posed even an even more difficult task.

By 1958 Dunlop engineers had designed a Maxaret unit for automobiles. Lister tested this unit on a Jaguar Mark VII with disc brakes on both front and rear. Making a Maxaret for automobiles was more than a process of miniaturization. The Maxaret for automobiles was no longer driven by a pneumatic wheel; instead, a geared ring inside the disc brake was in contact with the wheel rim. The exhaust valves could be opened without shutting off the master cylinder supply. A modulator valve allowed the brake to be pre-pressurized to allow the near contact of the shoes with the disc, meaning that the brake achieved full application almost immediately. Consequently more cycles of modulation occurred. Although this addressed the concern about low braking efficiency seen in the initial Maxaret trials, it actually caused a more significant problem. Because the brakes were engaged so strongly at first, they caused the car to shudder, leading to a potential loss of steering control as the driver lurched forward. This had not occurred during the first trials because the brakes did not engage quickly enough, although the delayed action was a problem in its own right.

On the plus side, sluggish steering was no longer a problem, as the unit permitted finer tuning and could distinguish between cornering and braking more effectively.[33] The vehicle's braking distance was tested on five different surfaces and at two different speeds on each surface. In all but one test, the Maxaret stopping distance was shorter than the locked-wheel distance, and stability was retained. Indicating his continued concern with the average quality of vehicles on the road, Lister also saw an advantage in the device's ability to correct for severe brake imbalance, whether it was due to misuse, poor design, or lax maintenance of the vehicle.

Lister's work on the Maxaret was widely published and presented at no fewer than a dozen conferences. Reaction to the device covered the spectrum between enthusiasm and dismissal. Most of the detractors simply did not believe that an antiskid device was needed; proper driving techniques were sufficient. Lister's experience at the RRL, where skidding was not seen as solvable through any one approach, was enough to convince him this was not true. Still, the Maxaret's problems proved to be even more difficult than they initially appeared. Vibration, including both the sharp, pre-pressurized actuation and subsequent jerking cycles, became a common problem in all attempts to transfer the aircraft-style brake to automobiles.[34] In addition, the Maxaret and other devices under development in the late 1950s and early 1960s all proved to be prohibitively expensive.[35] By the mid-1960s companies were developing new systems with an eye to affordability, and this led to a new set of compromises in quality and functionality. Still, the RRL's experiments with the Maxaret in the first decade of research on antiskid devices encouraged optimism about the possibility of an antiskid device, and this was a most important first step.

What the RRL's program did not produce were wide-reaching solutions to skidding problems. The road surface group realized that their great successes were small-scale projects to improve particularly bad patches of road, where accident rates were abnormally high. The driver interaction groups helped eliminate many potentially hazardous conditions, especially in publicizing the problems of alcohol and driving, but had no real effect in training drivers to avoid or better react to skidding. The tire design groups worked to keep skidding reduction in the minds of tire manufacturers, who tended to concentrate more on fuel efficiency and tire life, which were often at odds with antiskid tread designs.[36] The brake designers investigated the potential of brake modulating units, but ran into difficulty developing wide-ranging solutions to the skidding problem when they encountered problems with the antilock unit's interaction with the brakes. The RRL's attempts to reparse the skidding problem without losing sight of its interactive tendencies appeared to other engineers to be a successful model of what an agency should do. Yet it still did not produce any broadly applicable answers to the problem. The interactive dimension of the skidding problem proved intractable, even under wise management.

CYRIL GEORGE GILES AND THE PATENTING OF TESTING TECHNOLOGY BY A GOVERNMENT AGENCY

C. G. Giles was another RRL employee whose work played an important role in the development of antiskid devices. However, Giles's contributions were unexpected, considering that he was not working directly on the development of antiskid devices, as Lister was. Giles's work illustrates how posing the skidding issue as an interaction problem yielded direct and surprising progress in the development of antiskid devices.

In the 1950s Giles worked in the surface characteristics section of the RRL, and by 1963 he directed the human factors subsection. As a result he had experience in two areas of the RRL's interaction problem, neither of which was braking systems. At the Charlottesville conference in 1958 he wrote four papers, one on the topic of road-tire friction and rubber hysteresis,[37] two on methods to measure skid resistance, and one, co-authored with Barbara Sabey, on a statistical assessment of accidents in the United Kingdom. He also presented Lister's paper on investigations of the early Maxaret system. As a result, Giles played an important role in the early skidding research community, but the trajectory of his published work up until 1963 would not lead one to expect him to prepare a patent for an electronic circuit that would compare wheel speed with vehicle speed, one of the more intractable ABS problems of the early 1960s. Giles was neither a braking system designer nor an electrical engineer, but he had become intimately familiar with the testing equipment necessary for measuring the same parameters an antiskid braking system needed to measure. The breadth of his involvement with the interaction problem at the RRL was critical to antiskid research. Giles is also a model participant in the development of the knowledge community. He was involved in multiple problems concerning road safety and clearly served as a vector to move new knowledge between different groups with which he was engaging.

On 12 March 1962 Giles wrote an internal memo to G. A. B. McIvor, who had replaced Glanville as the director of the RRL: "I have recently seen patent specification No. 888,824 of 7 March 1960 for an antiskid braking system using electrical operation. For some time now I have been thinking of systems of this type and the problem of how to get a satisfactory reference voltage when the brakes are applied to all the wheels of the vehicle. Arising from the work on an automatically controlled car, a novel solution to this problem has recently occurred to me."[38] Giles then described his novel solution as something recognizable as a component of an antiskid device: the idea is that the brakes are fully or partially released whenever the output of the generator representing wheel speed falls below some reference voltage, usually taken as representing the speed of the vehicle over the road.[39]

Giles's experience in assessing the friction properties of different road surfaces is evident in his description of the problems he was concerned with solving: "Since many factors influence the adhesion between tyre and road surface, the retardation available in any emergency situation can vary within wide limits and it is not therefore possible to represent the deceleration of the vehicle with any accuracy by a reference voltage which falls in an arbitrary preset manner."[40] Giles's letter resulted in McIvor's pursuing a patent of the circuit, first through the patents section of the DSIR and later directly through the National Research and Development Corporation (NRDC), a corporation set up by the British government to facilitate the commercial development of products invented in government scientific laboratories. Initially H. A. Howe, of the DSIR's patent division, resisted pursuing the patent, since he doubted that the device was actually functional. He wrote, "We might have to have considerably more detail of how the invention could be put into practice."[41] Giles dismissed this concern by claiming that the circuit would come into use in the next phase of the work on the automatically controlled car for friction measurement he was undertaking for the RRL (project DS17).[42] He then invited Howe to come to Slough and take a look. McIvor's response was quite different, possibly reflecting his greater familiarity with the scope of projects being undertaken at the RRL. McIvor pushed Giles to discuss the possibility of combining his work on the voltage generator with Lister's brake proportioning system. He also requested that Howe be present for these discussions. But Giles seems to have ignored all of McIvor's suggestions for cooperation with Howe and Lister.[43]

Even without combining Giles's and Lister's projects, McIvor decided to proceed with patenting Giles's (as well as Lister's) inventions and sent all of the available information through the DSIR's patent office to the London patenting agency of Sefton, Jones, O'dell, and Stephens. Still, this did not occur without further questioning from the patent office at DSIR. C. J. P. Meade, a DSIR patent officer, requested additional information, noting that "the charging and discharging of a capacitor follows a logarithmic law, whereas a relationship between velocity and acceleration is a linear one."[44] Giles responded with a schematic and claimed, "The basic difficulty arose because neither Mr. Meade nor the Patent Agent were really familiar with the sort of analogue computing techniques on which the proposal was based. . . . [After explanation] Mr. Meade is now happy to go ahead."[45]

In September 1963, nearly a year later, Giles filed an internal patent application for the NRDC. He responded to a question about the scope of the patent by writing that the development was "a minor improvement at present time, [but] could become very important [if] the antiskid braking systems with electrical operation come into use on road vehicles."[46] He also specified, "Goodyear Aircraft Corporation has an antiskid braking system in which something of this kind could be used. . . . Goodyear's patent for the electric aircraft brake uses a condenser system to get its reference voltage. It was this which suggested the present alternative. Some people have even gone to the complication of Doppler Radar to achieve the same result."[47] Giles's research into antiskid systems for his automatically controlled brakes clearly led him to examine the leading publications of the aircraft braking community. Yet it also appears that he was not looking at the same devices as Lister was, since Lister's patenting approach focused to a greater extent on the aircraft antiskid systems produced by Lockheed and Perma.[48]

By the middle of 1964 the discussion between Giles and the patent agents focused on whether the reference voltage circuit should be patented in the United Kingdom alone, or whether U.S. patent protection should also be sought. Giles pressed consistently for U.S. patent protection. Although he never claimed that a market existed in the United States that justified his device's patent protection, he sought to discuss his findings with American researchers, especially those at Bendix, an American brake manufacturer. Giles sent a copy of a paper by K. V. Bailey of Bendix to A. F. Cooper, the head of the mechanical engineering group at NRDC: "I am enclosing a copy of an American paper on an electronic anti-skid brake system for road vehicles which has just appeared and you might find interesting. I think my device would be a distinct improvement on the functioning of equipment of this kind. . . . In view of this American development, you might like to reconsider the decision about patenting the idea in this country only."[49] Cooper's response was blunt; he advised Giles that there was insufficient hardware to demonstrate to manufacturers, and that since the project was not a priority in Giles's department (the road surface department) there would likely be no forthcoming prototype. The antiskid device remained a testing instrument for Giles and his investigators, unlike Lister's involvement with the Maxaret, which was a collaborative development project unto itself. Giles wanted U.S. investigators to have access to his circuit for testing machines, not solely for its own sake as a component of a braking system. Yet this status as a component of a testing machine put Giles's circuit in a precarious and nonprioritized position. So, against the advice of Sefton, Jones, O'dell, and Stephens, Giles publicized the circuit.[50] This technology was, for Giles and the RRL, a means to an end.

Whereas others at the RRL faced the complexity of the interaction problem unsuccessfully, Giles was able to use the resources of the RRL to do work which benefited from his familiarity with devices for measuring skidding as well as the RRL's interest in preventing skidding. The RRL provided an ideal location for a testing instrument to become a component; it offered an environment in which Giles, an engineer who spent much of his time measuring friction, could read articles in electronics and aerospace journals. Giles applied this information to increase the accuracy of machines that measured road surface friction and in the process solve a problem which had stumped braking specialists for a decade. His hands-on experience and knowledge of measurement provided the spark.

By 1962 Giles's work on the reference voltage generator had solved one lingering problem of antilock braking systems. All the antilock systems developed in the 1960s needed to generate a reference voltage with which to compare the wheel speeds, which could also be expressed as a voltage. Giles's circuitry became prototypical, and because he made the design public, all members of the antiskid community had access to it. Giles's work, unexpected for a civil engineer engaged in measuring road friction, had set the stage for the development of electrical and electronic signal processing. However, Giles continued to work as a road engineer at the RRL, and the RRL was not in the business of building braking systems. Consequently, for Giles's work to have any legacy, it had to be adopted by the different parts of the antiskid community. Giles's circuit added to the cumulative knowledge of the community through the work of engineers whose goals were to design antilock systems. These developments in the private sector are the subjects of the next several chapters.

THE INTERACTION PROBLEM AND THE BRITISH INDUSTRIAL RESEARCH ORGANIZATION

Many historians have discussed the unique structure of British industrial research in the twentieth century. Most often, the British case is used as a comparison with the American, West German, and Japanese systems to show that underinvestment by industry in scientific and engineering research leads to poor rates of innovation and a relative decline in industrial and economic power.[51] The British system of industrial research focused on government oversight and support by the DSIR for research performed at government research centers, such as the National Physical Laboratory and the RRL. The government also supported research at industry collective organizations, such as the Motor Industry Research Association (MIRA), formed in 1946.[52] The purpose of the research associations was to investigate problems common to all members of an industry. Often the research associations took on standardization and specification work, and frequently their work in many different industries fell into the category of materials science. Projects undertaken by MIRA in 1946 included the chemical estimations of the wear of steel shafts, the development of strain gauges, the development of a low-frequency oscillator, and research on the performance of exhaust valves using leaded fuel.[53] Because the RRL had established an interest in skidding when MIRA was in its infancy, there was no demand for MIRA to contract to perform skidding research. Simply put, MIRA did not engage in skidding research because the RRL did.

In fact, the skidding problem was uniquely suited to the British system of government-sponsored research. The RRL possessed both human and physical resources that even the world's largest corporation, General Motors, could never put together. In addition, the team approach to research that was frequently unsuccessful elsewhere worked well on the skidding problem. An article in *The Engineer* in 1950 explained the British approach to engineering management:

It is not the purpose of this paper to discuss the process of engineering development from idea to finished design, but it is perhaps desirable to record the minimum facilities which it demands. These, in the order of their application, are:

1. A research department which forms a reservoir of knowledge, collects new knowledge by new investigations, and converts it into a form which the design department can use.

2. A materials laboratory which advises on materials for new projects and (at a later stage) prepares material and process specifications. . . .

3. A design department (drawing office is too narrow a term). On the desks and drawing boards of this department all the other work of research and development finally converges and is converted into drawings and specifications.

4. An experimental office in which new products are built and tested.

5. The inspection department, which supervises all inspection outside the province of the materials laboratory. . . .

Coordination implies management, and in the skeleton of an engineering department, as indicated above, there is a significant change from the earlier view of an individualist chief engineer or chief designer, towards the more objective and analytical approach suggested by the term "engineering manager." It does not and cannot replace creative thinking, but aims to supplement it by greater attention to detail and by a greater degree of specialization within the engineering department as a whole.[54]

The organization of projects at the RRL followed this schematic very closely. The division of intellectual labor provided an entry into complicated interactive problems such as skidding that an individual chief engineer could not fathom.

Despite the effectiveness of the RRL in dividing the skidding problem into more manageable units, Rosenberg and Mowery's claim that the British system was unable to transfer knowledge provided by the government research organizations to industry holds true in the case of skidding. As early as the 1920s, a governmental committee headed by Lord Balfour was critical of industry's ability to receive and apply new knowledge.[55] Because companies rarely hired engineers or technologists to exploit extramural research, research produced by the research associations and government laboratories was often inaccessible to industry.[56] In the case of skidding, Dunlop and Jaguar were recipients of the RRL's research, as well as collaborators in it, but the cost and complexity of the Maxaret device was unattractive to the vast majority of automobile manufacturers. The fact that the Dunlop Maxaret system appeared as an option on a 1966 Rolls-Royce and a 1968 Jensen was little more than a blip on the radar screen of antiskid systems.[57]

The work of the RRL on the skidding problem occurred in a unique cultural window of opportunity. The early 1950s were situated between the years of significant enthusiasm for engineering which immediately followed World War II and the sober recognition of Britain's declining world economic and industrial status by the late 1950s, followed by the "stagnant society" of the 1960s.[58] An editorial written in 1955 in the British engineering weekly, *The Engineer*, reported on the blinding effect of the "curious optimism that appears to be bred by war experience."[59]

In the late 1940s British engineers displayed an optimism about many things which had been intractable.[60] Difficult problems deserved further investigation, and sometimes solutions were just an experiment away. Britain's scientific prowess had been borne out by her performance in the war. *The Engineer*'s "Retrospective" for 1945 evidenced this attitude:

The fact remains that our native science was able to meet every call that was made upon it, and that our industrialists were able to translate rapidly into workaday things the inventions and discoveries of scientists. . . . It is sufficient in this place to insist that throughout the war British science and industry—we make them one—refound itself. With that lesson before us we can enter the future years in the sure knowledge that keeping confidence in ourselves, we are competent to do all that we may be called upon to do.[61]

During the two world wars British engineers earned great credibility for themselves and their profession grew in part because of their visibility: "The Institution of Mechanical Engineers has been rapidly growing in membership for many years. Moreover, as a result of its experience in two world wars and more particularly in that just concluded—an engineer's war, if there ever was one—it has achieved a 'national' rather than a purely 'technical' importance."[62]

The postwar world was promising to engineers because of their success in the war. Research which had stalled before World War II was tackled in the postwar period. These problems were both scientific, bolstered by the splitting of the atom,[63] and socioeconomic, exemplified by the construction of the welfare state and the global monetary system developed at Bretton Woods. Industry was expected to outperform its interwar record as well, leaving those disasters as anomalies in the glorious history of Britain:

Industry must make a productive effort substantially greater than 1938 but with approximately the same manpower. That could only be achieved by a broad measure of cooperation between Government, employers, and operatives and by the most efficient use of scientific research in its wider sense and throughout the widest field. The aim must be a new industrial revolution in this country, and we must contrive in the years ahead to reproduce the teamwork, the close identity of purpose and action between science, management and labour which was the outstanding feature of the war effort.[64]

By 1960 engineers knew that a new industrial revolution was not going to occur in Britain. The mood of the late 1940s was referred to as "curious optimism," implying that this optimism was unfounded. Social, economic, and political troubles had arisen, and engineering had not provided solutions. Economic growth was sluggish after the war and nonexistent by the mid-1950s.[65] T. O. Lloyd writes in his history of twentieth-century Britain, *Empire, Welfare State, Europe*, "Suez and the worsening of industrial relations were symptoms of a feeling that the post war world was less pleasant than had been thought."[66] The Two Cultures debate in the late 1950s, beginning with C. P. Snow's article in *The New Statesman*, was a poignant statement of the backwardness of England in the 1950s, yet it failed to correct any problems of engineering or scientific education. Kenneth Morgan saw Snow's theme as the inability of an aging society to adapt itself to the challenge of the new. Regarding the mood in 1963, Morgan wrote:

The current of opinion was that British skill at invention and scientific pioneering had not been followed up by the effective application, especially to the manufacturing industry. Britain since 1945 had pioneered atomic energy and produced the world's first nuclear power stations. It had developed jet aircraft and maritime propulsion; it had led the way in the development of computers. But now it was being taken over by others. . . . The gap, not only with America, but with Western Germany, France or Sweden grew even wider. Industrialists were vocal in the need for wider investment in science, in place of the "big science" projects of the fifties.[67]

It was just this environment of big science in which the RRL's investigation of the interactive problem of skidding thrived. When that environment no longer existed, Britain's leadership in the problem of skidding ended.

In the 1960s the RRL was losing its primacy as a model for the organization of research, particularly for skidding. The work of Lister in particular had served to generate interest from the private sector in the potential of developing a marketable antiskid system. The notion of skidding as an interaction problem had served its purpose as an organizational trope, but after a decade of research the problem looked different and different actors and agents were prepared to play new roles in further research. The DSIR was attracting less funding from Parliament throughout the 1960s, and the RRL attracted less DSIR money. The RRL focused more closely on government works projects, and space for R&D projects like Lister's and Giles's vanished.[68] Government sponsorship of research appeared to be a failure and the British government was cutting back.[69] Companies with an economic interest in the solutions to skidding would have to commit to performing their own research. These corporate interests split research efforts in new ways, with the R&D on tires and braking systems projects having little cross-fertilization. To be sure, individuals, including Lister, remained important to the development of ABS, but the RRL and Britain lost their position with the declining industrial clout of Britain. While the RRL produced both important ideas and critical machines for skidding research, their model of a division of labor did not produce antilock braking systems per se. By the late 1960s their failure to produce a commercially viable system also defined the RRL. In other words, a different type of organization was needed to move antilock braking system R&D to the next phase.

four

The Track and the Lab

Brake Testing from Dynamometers to Simulations

R. D. Lister and C. G. Giles constructed knowledge in the process of designing machines to measure the performance of
braking systems at the RRL. They began the process of transforming their knowledge of one-off prototypes of measuring
machines into devices that could be produced commercially to
prevent the wheels of a car from locking up. However, neither
Lister nor Giles was working directly on the problem of building a commercial ABS. Producing commercial technologies was
not what the RRL did, after all. Still, Lister's and Giles's designs
of machines formed the backbone of knowledge about skid
prevention devices in the early 1960s. While they moved remarkably easily between the public and private sectors, others
who were interested in improving the skidding performance of
braking and tire systems worked exclusively in the commercial
sector. Lister's and Giles's work promised that antiskid devices
could be commercially viable. Studying skidding was no longer
limited to public research agencies such as the RRL and the
Virginia Department of Road Design but was moving into the
private sector. Transforming skidding from the intractable "interaction problem" Glanville had proposed at the RRL to a
problem preventable by a new braking system made the problem of skidding attractive to the commercial sector. Nevertheless, the mere fact that antiskid devices were commercially
promising did not immediately yield solutions; corporate de-
sire alone did not solve technical difficulties, which ABS was
fraught with in its early years. However, thinking practically

about making viable devices for skid prevention did bring new practitioners into the community, and they once again changed the way skid prevention was defined. Redefining the problem centered on measurement and testing, and engineers specifically asked "What shall we measure?" and implicitly asked how.[1] The questions of *what* to measure and *how* to measure it were inseparable once again.

The essential issue in the development of measuring and testing braking systems consisted of translating the qualities of handling into measurable quantities. Lister's work changed the way engineers involved with measuring the performance of a braking system thought about the qualities they were trying to quantify. In the early 1950s, prior to Lister's dominance, discussions of skidding often circled around intangible qualities such as "feel for the road," "road grip," and "effective handling." But how was an intangible quality such as "feel" measured other than by surveying expert test drivers? How could one be assured that each test driver gave the same verbal rating to the same feel? William Thomson's famous quote seems particularly appropriate here: "When you can measure what you are speaking about and express it in numbers you know something about it; but when you cannot measure it . . . you have scarcely in your thoughts advanced to the stage of science."[2] In other words, measuring was no more a pure idea than a machine was. The process of learning how to measure required making measurements, just as the process of design required producing designs.

In the case of metrology, until the community of practitioners negotiated how they would quantify certain qualities, they had no way to be sure they were discussing the same qualities or the same standards. The process of making quantitative determinations of handling qualities contained two components: negotiating a common verbal understanding of these esoteric qualities and correlating measurable quantities to these definitions. This process required engineers familiar with the two main loci of automotive testing: the track and the lab. The development of methods to measure skidding in the 1960s pitted these two locations against each other, with the ultimate resolution in the 1970s that both were necessary. Still, by the end of this period, computer simulation threatened to replace actual testing altogether.

Even after World War II automotive engineers familiar with the track thought of the RRL's 1931 rigid wagon as the absurd state to which road testing would be doomed.[3] Testing models was inaccurate and overly simplified, but full-scale road testing ended up being physically unpleasant and dangerous for many engineers, who were often also the test drivers. The RRL and other engineering societies strove to eliminate the difficulties of road testing with improved bench tests. The Society of Automotive Engineers (SAE) in the United States played a major role in the effort to replace road testing. In 1951 the SAE commissioned a group of engineers representing eleven companies to create Brake Sub-Committee 3. The SAE gave this committee the mission of producing a code for standardizing the use of inertia dynamometers in brake testing. Sub-Committee 3 produced the report "Construction and Operation of Brake Testing Dynamometers," which was presented at the annual meeting of the SAE in 1953.

Dynamometers made up the largest group of bench testing instruments used in braking research. The term *bench* in this context is misleading, since these instruments were far too large to sit on a bench. The term was used simply to contrast laboratory testing instruments with tools and machines used to measure the performance of a vehicle on the road or test track. Dynamometers had become common instruments in the development and maintenance of automobiles in the postwar world. Engineers involved in the work of designing dynamometers for automotive use even named their work *dynamometry*, by which they meant "the art of applying a dynamometer to the problems of automotive service work."[4] The SAE report of 1953 described the state of the art in the early 1950s. These engineers championed bench testing as a more simple and elegant alternative to the expensive, messy, inaccurate, and often dangerous work of road testing. For the engineers who produced this report, road testing meant losing control of the conditions of testing, such as temperature, speed, and load. Experiments performed under carefully controlled conditions could be replicated more easily, which made the results more useful for the sake of comparison. As the main purpose of the inertia dynamometer was to compare the performance of different designs, duplicating test conditions proved to be of the utmost importance. The authors of the SAE report wrote:

The inertia dynamometer is invaluable in permitting observation and recording —in many stages—of the pertinent characteristics of a vehicle brake which indicate its probable performance on the vehicle. In view of the fact that the inertia dynamometer cannot duplicate all road conditions or complications resulting from the installation of the brakes on a vehicle, it is limited to the role of screening or providing data for comparisons of brake operating characteristics measurable within its scope. The inertia machine cannot eliminate the test work required to evaluate brakes on the vehicle, but it can supplement and shorten the amount of road testing necessary.[5]

In short, when engineers defined their projects in ways that made replicable experiments and similar testing conditions critical, they constructed arguments in favor of bench testing. When duplicating the unpredictable nature of the street became their goal, bench testing's only advantage consisted of its relatively lower cost. Engineers notoriously resented choosing a less satisfactory testing method for the sole benefit of cost. Furthermore, simply being less costly than road testing did not mean that these instruments were cheap. While they were not expensive to operate, these inertia dynamometers cost up to $150,000 and the largest had flywheels over five feet in diameter and weighing 15,000 pounds.[6] Clearly these machines were designed for large, wealthy corporations or research agencies.

The inertia dynamometer provided a measurement of torque, a variable that had proven difficult to measure directly in road tests. Brake torque was the primary comparison used in the design of braking systems in the 1950s. However, researchers working on skidding did not attach the same significance to this variable, preferring to measure and compare rates of deceleration.[7] In addition, the users of the inertia dynamometers of the 1950s made one assumption fatal for researchers interested in skidding: they assumed that all the wheels decelerated at the same rate. There was only one flywheel, so differences in the rates of deceleration between the wheels could not be determined. Because inertia dynamometers tested one brake at a time, braking system balance and distribution also could not be determined. Moreover, the machines of the 1950s were too large to use more than one flywheel and still fit all of a vehicle's wheels on it.

In the early 1960s engineers at Johns-Manville presented their design for a new, more sophisticated dynamometer. David Sinclair and W. F.

Gulick published their paper on the dual brake dynamometer in 1963 as an improvement on the inertia dynamometers of the 1950s, claiming that measurements made on their machine correlated more closely with actual road tests.[8] Sinclair's and Gulick's research centered on the relationship between torque transfer and brake fade. Instruments in the vehicle measured stopping distance, velocity, deceleration, line pressure, and, most critically for Johns-Manville, temperature.[9] Sinclair and Gulick were also able "to introduce several random variables such as wind, ambient temperature, humidity, dust, car tilt, rolling resistance, and windage."[10] However, before their machine could accurately duplicate road conditions, they needed to know more about the behavior of a car on the road. Most measurements, even those of the RRL, were aggregate measurements, such as average deceleration, rather than instantaneous deceleration at the most critical moments. Sinclair and Gulick ended their paper with two key observations: that more had to be known about deceleration and velocity in order to make the dynamometer mimic road conditions and that their real goal was a four-brake dynamometer.[11]

For research in skidding, dynamometers had other clear drawbacks. The subcommittee admitted, "Considerable judgment and experience are required to predict from dynamometer test results the actual vehicle performance involving two or more brakes."[12] For the most part, engineers investigating skidding were looking at unpredictable behavior, so the need to extrapolate quantities from measurements made on inertia dynamometers proved to be a real weakness. Because the engineers had to make so many assumptions about the way dynamometer readings might relate to measurements taken on the road, the results of their work were rarely surprising. In contrast, skidding occurred suddenly and unpredictably. The single-flywheel inertia dynamometer proved useful for research into the frictional surface of brake drums and shoes, but engineers were never able to use it to provide information which predicted the behavior of an actual automobile. The control over experimental conditions that the dynamometer supporters championed meant that the dynamometer could never provide the information that skidding research needed.

Bench testing could improve if engineers knew more accurately how to simulate unpredictable road conditions in the laboratory. But despite the utility of the dynamometer as a tool to test and evaluate brake linings, it was the wrong tool for examining skidding. Investigations into skidding required real-time, real-space testing of actual vehicles, not bench testing. The emphasis on more accurate instruments for road testing focused on providing highly accurate readings of particular variables at particular points in the braking process. While civil engineers working to design higher friction road surfaces had not yet achieved this level of accuracy, their tools were increasingly designed to provide measurements of change over time, exactly the parameter skidding researchers were seeking. The use of the wrong tool for measuring skidding led engineers early on to comprehend and appreciate the complexity of skidding. Measuring one wheel and extrapolating the behavior of the others would not be good enough. Real time, space, and conditions mattered when it came to skidding; laboratory simulations proved deeply unsatisfactory.

The dynamometer, which had seemed to be one of the keys to designing better braking systems in 1950, was found incapable of quantifying skidding behavior by 1960. Nevertheless, bench testing had been an important domain for engineers in the development of improved braking systems. Learning the shortcomings of bench systems was an important step in developing better machines to measure performance outside the laboratory. As the next stage, field testing proved to be far more complex and ultimately more important to the development of antiskid devices. Consequently, a different group of engineers asked a different set of questions in the process of designing instruments that could better measure the behavior of a skidding car in the unpredictable conditions of road tests. The effort to replace expensive and complicated road testing with bench tests distinguished the use of research technology in the early 1950s from what superseded it. By the late 1950s the effort had switched to making machines that made road testing less dangerous and expensive and more precise.

Even as some engineers worked to improve the performance of the large-scale dynamometers in the 1950s, others increased their commitment to road testing. The work of the RRL on the fifth wheel apparatus provides an example of this. But posing the question of how best to test brakes as a battle between road testing and bench testing is an over-simplification. To many instrument makers the issue of road testing versus bench testing was not an either-or conundrum; whether measuring instruments were mounted on a bench or a vehicle, much of the technology was generically applicable. In fact, the majority of instruments designed to test braking systems were not tailored specifically to measure skidding. In the 1950s and 1960s automobile and brake manufacturers developed a great deal of new braking technology. A significant amount of this technology was critical to the subsequent development of adaptive control technology, including ABS. These engineers formed the diverse core of the ABS community, as they came together from the automobile industry, the tire industry, and government agencies such as the RRL and the Virginia Council of Highway Investigation.

ENDING THE DEBATE OVER BRAKING SYSTEMS

In the 1950s and 1960s auto manufacturers asked what the limits were of braking systems performance. While engineers working on high performance and racing vehicles primarily focused on this question, eventually it trickled down to production line vehicles. Most often, engineers wondering how to improve the overall performance of a production braking system engaged in a debate over whether disc or drum brakes were better. Disc brakes required higher hydraulic pressure and were more mechanically complicated; therefore they were more expensive, and their added cost had to be justified by their performance. They were standard on high-performance sports cars and racing cars as early as the late 1940s.[13] A quick survey of the annual brake reviews in *Automobile Engineer*, a British publication, highlights the developing preference, especially in Europe, for disc brakes. *Automobile Engineer* published these reviews immediately after the annual international auto show in Geneva, for years the largest in the world. In 1958 the review offered this assessment:

Of particular interest is the increase in the number of cars exhibited with disc brakes, applied either to all four wheels or to the front wheels only. The future prospects of this development are not easily assessed and are still largely dependent on economic considerations, particularly as far as mass-produced cars are concerned. As in the case of commercial vehicles, in spite of certain lingering defects, the advantages of disc brakes have been well established, and have earned for this type of brake unquestioned acceptance for many racing and sports cars. There is, however, a considerable weight of opinion in support of the argument that drum brakes of current designs adequately meet the safety requirements of ordinary motoring. In view of the present high level of purchase prices, largely because of taxation, there is no great incentive to incur the additional cost of disc brakes as standard equipment and still less as an optional extra.[14]

By the auto show in 1965, a major shift had occurred in manufacturers' attitudes toward disc brakes. No longer were they prohibitively expensive, and drum brakes were increasingly viewed as substandard. Between 1963 and 1965 the use of front-wheel disc brakes became increasingly common even in cars made by the three major American automobile manufacturers. The editors of *Automobile Engineer* found this swing to be "perhaps the most significant of all brake changes in recent years."[15]

The increasing popularity of disc brake systems in the 1960s reflects the effort of brake manufacturers to make these systems less expensive. It also indicates a greater knowledge of the advantages of disc brakes, as well as the ability of the automobile industry to sell safety features. The marketing of auto safety features began long before Ralph Nader brought the safety of automobiles into sharp focus with his exposé *Unsafe at Any Speed*, published in 1966. Whether the disc brake's safety was being sold to the public or to the automobile manufacturers, the issue required extensive testing and measurement.

The West German Verein Deutscher Ingenieure (VDI, Association of German Engineers) performed the most extensive comparison of drum and disc braking. The results of their tests were published in 1962 in the *Automobiltechnische Zeitschrift*.[16] The VDI relied on road tests as well as dynamometer tests. For the road tests, they used a near copy of the RRL's fifth wheel apparatus. They compared dozens of variables but focused on the disc brake. One of the primary difficulties that must be overcome in the design of disc brake systems is the heat that is generated. The VDI did not intend to produce a report claiming one system's superiority over the other; this would have violated the commercial impartiality of an engineering society. Rather, the report produced by the VDI, authored by Bielecke and Bethke, who were employed by Jurid Werke, GmbH, a machine tool producer in Hamburg, aimed to set a standard for the type of information companies needed in order to choose one design over the other. The VDI created a chart of the variables which engineers should measure. How these quantities were used in making choices formed a different question altogether, one which the VDI had no intention of answering. Still, their standards for testing aided in the adoption of disc brake systems, particularly on the front wheels of a car.

Since early ABS developed in tandem with disc brakes at places already working on marketable systems, such as Dunlop, developments that improved disc brakes also improved ABS. Nevertheless, the VDI facilitated communication within both the brake testing and ABS communities as they matured. The success of auto manufacturers in getting consumers to pay more for safer cars also contributed to the commercial interest in what promised to be expensive antiskid mechanisms. Engineers involved in the development of disc brakes often faced the same difficulties as engineers working on ABS would a decade later. The long development process and eventual success of production disc brakes made long-term R&D projects related to safety issues somewhat more promising to corporations and provided some companies with experience in seeing an expensive development project through two and a half decades of preparation.

Combining road and bench tests to define quantifiable parameters of brake performance was the primary innovation of the VDI tests. Still, engineers interested in how the brakes affected the handling and performance of the vehicle could not switch back and forth between laboratory- and track-derived quantities. They needed a set of track-based instruments that could work in real time and in the space of a standard passenger auto. These instruments also faced the frequently harsh conditions of the track, in terms of both weather and vibration. Testing machines that involved large trailers modified the handling characteristics of a vehicle and therefore defeated the purpose of track testing. The universal friction tester built by the mechanical engineering program at the Technische Universität at Stuttgart in the 1960s was contained in a trailer nearly the size of a semi. Before World War II engineers knew that a vehicle's propensity to skid varied according to the weight of the vehicle, even if all other factors were equal. Adding a several-ton trailer clearly produced side effects that did not scale linearly from one size vehicle to another; in this case, there was no consistent mathematical relationship between the likelihood of a tractor-trailer and a passenger automobile skidding. As a result making an automotive system involved more than simply proportionally miniaturizing the same system the trailer used. Not only was scaling the instrument's behavior complicated, but the instruments had to fit in the completely different physical space of a passenger car, which needed to accommodate a driver, passengers, and cargo. This size constraint had a controlling effect on the technology engineers developed in the 1960s. In addition, placing the measuring system in a car required a degree of robustness unknown in the lab. Along with space constraints the vehicle-mounted system faced vibration, temperature extremes, dirty conditions, and extremes of moisture. While engineers designed cars that could brake automatically, the presence of a trained test driver also became an important part of the process of moving from qualitative to quantitative assessments. Ultimately, test drivers proved more reliable at duplicating stops on a track than machines; the development of road testing failed when engineers tried to discard the driver and succeeded when the driver was included as part of the system. Engineers had to incorporate the interaction problem into the way they thought about braking.

THE FORMATION OF THE COMMUNITY IN BRAKING SYSTEM RESEARCH TECHNOLOGY

Terry Shinn writes about a category of scientists and engineers he calls research-technologists.[17] Explained briefly, research-technologists are the engineers, scientists, and technicians devoted to the design, construction, and application of new metrological tools. Their research focuses on developing these tools and their protocols, but not necessarily on the use of the tools in specific scientific environments. So research technologies are designed to move between different knowledge communities, but they often bring the imprint of their original use and development with them. The use and development of metrological instruments by the ABS knowledge community is a good example of the movement of instruments and how they bring certain assumptions with them.

For better instruments that provided numerical references to the feel of the car, automotive engineers looked to the aircraft industries, as Lister had in his initial experiments with the Maxaret. The airplane designers had long wrestled with ways of translating intangible qualities into numerical equivalents.[18] The designers of systems for the control and handling of planes often subscribed to the maxim that control and stability were inversely proportional; that is, the more control a pilot had, the less inherently stable a plane's design was, and the more stable an airplane was in the air, the less control a pilot had. Because of this peculiar quality, as well as the ongoing struggle between pilots and engineers over ownership of these design decisions, the handling qualities of planes had been scrutinized with the goal of quantifying this relationship since before World War II. The process also established an important role for pilots in determining quantitative correlations to the subjective qualities of airplane handling.

Automobile designers looked to aeronautical engineers for guidance in both the construction of theories of vehicle dynamics, often modified from theories of aerodynamics, and the design of instruments which would allow them to establish numerical correspondents to the intangible qualities of handling. Both Edwin Layton and Edward W. Constant have presented other cases where instruments formed a middle ground between theoretical, scientific knowledge and engineering design.[19] Instead of seeing the instrument as a nexus where science and engineering came together, engineers developing machines for measuring vehicle handling effaced the differences between scientific knowledge and engineering practice. More important, machines for measuring the qualities of skidding brought together the public agencies devoted to documenting the problems on roads with the companies who would fund the development of commercial antiskid devices. Where theory met machinery, the public and private sectors were bound to meet eventually.

The Cornell Aeronautical Laboratory in Buffalo, New York, emerged as a critical location for the modification of theories of airplane dynamics to vehicle dynamics. William Milliken and David Whitcomb led the investigation into transforming mathematical models of airplane stability into mathematical models of a car's handling characteristics.[20] Because the construction of these theories depended on a significant amount of new quantitative information about the dynamic behavior of an automobile, experimental data became a crucial input in creating new theories pairing the parameters of motor vehicles with models taken from aeronautical design. Engineers constructing these theories, most notably Milliken and Whitcomb, spotted an immediate need for new empirical knowledge. Consequently, engineers at the Cornell Aeronautical Lab also worked to design the new instruments needed to provide these data. The engineers producing these instruments were not imports from aerospace engineering; they were designing braking systems. The network of technical people working on the problem of skidding was becoming larger, more multidisciplinary, and represented more areas of industry.

One engineer who played a particularly important role in creating machines to measure the intangible qualities of automobile handling was Jean Odier of s.a. Française Ferodo in Paris. Ferodo, with both British and French divisions, was the world's largest commercial researcher into friction in the 1950s and 1960s. In an article that highlighted Ferodo's new English research facility that opened in 1959, the company's research areas were described as follows: "fundamental research on friction and raw materials; the development of improved types of brake linings; the testing of such new linings in the laboratory and on the road; and experimental production."[21] Odier, a graduate of the École des Mines, became the director of research in the physical and chemical division by 1962. His early work was highly theoretical, but this experience led him to yearn for better correlation between theoretical models and measurable results obtained from the test track.[22] Consequently, he turned his attention to building machines to make these measurements in the 1960s. Odier's primary contributions to the development of antiskid devices consisted of his efforts in two areas: building systems to simulate road conditions in order to facilitate more accurate measurements and his use of the results of these tests to construct mathematical models of vehicle dynamics. Odier wanted to create machines for reproducing in the laboratory the conditions of the road. Unlike the engineers who worked in dynamometry in the 1940s and early 1950s, he and his colleagues were not trying to eliminate the messy, unpredictable aspects of road testing, but rather to replicate this kind of uncertainty in the laboratory. Doing this work in the laboratory, instead of on a testing track, had several advantages; paramount among them were improvements in cost, convenience, and safety.

Odier's contribution to testing rigs consisted of a dynamometer which could simulate high speed, high deceleration, and high centrifugal force. Recorders at each wheel measured and recorded graphically several variables. Displacement, speed, acceleration, deceleration, and torque were detected at every wheel, while vibration, reaction of the ground, pivoting torque, braking ratio, driver corrections, and air resistance could be determined for the vehicle as a unit.[23] The dynamometer allowed engineers to run tests that would have been prohibited because of the danger to a human driver and provided a more solid theoretical basis for designing braking systems. Ferodo's dynamometer was available for rent to other brake-producing companies, and engineers constructing theories of vehicle dynamics, such as those at the Cornell Aeronautical Laboratory, used the results of Odier's work to set their parameters and provide test variables. The Cornell Aeronautical Laboratory relied on the results of Odier's work to construct the most elaborate theories of vehicle dynamics. Odier became one of the primary contributors to the translation of qualitative effects into measurable quantities useful in the construction of vehicle dynamics. One could argue that his oversight of the construction of the simulation dynamometer had a far greater effect than did his or anyone else's theoretical work. Once again, the design of the machine mattered the most. Being able to simulate dynamic behavior helped develop systems that could modulate their response instantaneously to changing road conditions. Odier's system functioned in real time, although not real space.

Engineers working on the problem of measuring handling qualities have stated the problem they faced: "There are two distinct sides of the objective measurement of vehicle handling: the first is the problem of providing the equipment needed to make the measurements; and the second is the problem of which measurements to make. The second of these involves not only which variables to measure but also the decision on what tests or manoeuvers to perform so that the measured movement of the vehicle during these tests can be used to describe its handling."[24] Engineers at the Highway Safety Research Institute at the University of Michigan worried about these issues of measurement. The Highway Safety Research Institute was another location for the development of theories of vehicle dynamics and the machines needed to provide the measurements on which these theories depended. Leonard Segel had worked with Milliken and Whitcomb at the Cornell Aeronautical Laboratory in the 1950s on translating theories of aerodynamic stability into vehicle dynamics.[25] In the early 1960s he joined the Highway Safety Research Institute, where he worked with Ray Murphy. In a career trajectory similar to Odier's, Segel and Murphy turned from building theories to building machines to provide the measurements their theories required. The career paths of Odier and Segel further reinforce Edward Constant's claim that "testing hardware and procedures embody substantial, technologically relevant, esoteric scientific information and therefore constitute a major mechanism for science-technology interaction."[26] While the notion of artifacts embodying knowledge is problematic, Constant's claim that testing technologies mediate between the relevant and the esoteric applies to the development of instruments to measure skidding and the construction of theories of vehicle dynamics using the measurements provided by these instruments.

In the mid-1960s Segel and Murphy modified an existing car to have a variable braking system.[27] This system allowed the driver to experience many different types of braking systems in a single vehicle. The system also possessed various recorders for correlating the test driver's qualitative assessments with graphical or numerical readings provided by the vehicle. Segel and Murphy envisioned the variable braking vehicle as a "research tool which will supplement analytical studies of the braking process. The vehicle will serve to validate theoretical predictions of braking performance for a large variety of braking system configurations."[28] They oriented their investigations in a theoretical direction, largely because this was the mission of the Highway Safety Research Institute. However, the testing machines they invented and publicized were widely used to establish quantitative tests for the qualities that braking system designers were seeking in the 1960s. Furthermore, unlike Odier's attempt to replicate road behavior in the laboratory, their machine was for use on the road. The variable braking vehicle was a simulator of a very different kind, one that simulated a specific type of braking system, not a road condition.

Perhaps the most important design aspect of the variable braking vehicle was the use of signal processors to simulate the behavior of different braking systems. This signal processor was an analog computer, sized to fit into the folded-down rear seat of a station wagon. This black box took an electrical signal generated by the driver's foot on the brake pedal and could process it to deliver a different amount or pattern of torque to a wheel. The variable braking vehicle was used by several different automobile, tire, and brake manufacturers in the development of new braking systems and tire tread and composition designs. But the more important role played by the variable braking vehicle was the precedent it set for inserting a computer between the action of the driver and the response of the braking system. Segel and Murphy worked this problem out in real time and real space, creating a system which could modify its response based on algorithms which defined the salient characteristics of particular braking systems. This system became a model for the designers of antiskid devices. Engineers were learning from one-off prototypes like the variable braking vehicle how to plan the design of an antilock braking system with electronic control.

Only a few miles down the road from the University of Michigan lay the General Motors Proving Grounds. At GM, G. W. Landon worked on another brake testing instrument, one that aimed to replace the inconsistencies of the test driver. Landon designed his Model 10 Brake Test Instrument to provide a controlled force on the brake pedal of a vehicle in order to test the braking system repeatedly under identical inputs from the driver.[29] The Model 10 allowed the tester to choose from four modes of operation: constant force, constant pressure, pulsing application of the brakes, or fully locked brakes. The Model 10 also contained equipment to plot force, pedal travel, deceleration, pressure, and time. The device coexisted with a driver who could override it at any time. Like the variable braking vehicle, this system played a critical role in the development of ABS, as the Model 10 had to perform many of the same measurements that a viable ABS would, and it operated under the same time and space constraints. However, Landon's effort to replace the inconstant driver did not help to correlate the qualitative and quantitative assessments of the vehicle's handling.

None of the machines discussed thus far explicitly addressed the interaction of the driver with the car. It was almost as though the work of the RRL had become lost in the effort to design better testing machines. For a test driver to play a useful role, he had to offer a precise assessment, one that proved consistent over repeated tests. Engineers remained skeptical that human beings could provide this service, yet only a skilled, trained driver could produce qualitative statements about an automobile's handling. MIRA in Britain produced a report in 1967 emphasizing the role that human qualitative assessment had to play in designing handling systems. F. D. Hales, N. F. Barter, and R. J. Oliver, all full-time employees of MIRA, wrote explicitly of the role the driver had to play, regardless of the sophistication of the handling system.

> The driver is an active component of the handling system, able to control the vehicle and its motions so that, subject to his and the vehicle's limitations, the vehicle takes on a desired motion. The driver himself acts as an adaptive control system, modifying his own behavior until the overall response is acceptable. The adaptation process takes time; how long depends on the ability of the driver. If a vehicle changes its characteristics significantly under some operating conditions, difficulties of adaptation may remain latent, emerging perhaps under critical conditions and resulting in a loss of vehicle control. Ride and handling are therefore in part psychological qualities concerning the subjective experience that an occupant feels as he rides in or drives a vehicle.[30]

Hales, Barter and Oliver addressed the problem of measuring these psychological qualities, writing that measurement was a two-pronged problem: designing the right equipment and determining what tests to use to measure and describe car handling. More than this, engineers had yet to find a way to correlate numbers being produced at the track with the human assessments of the qualities engineers knew were important.

At Ferodo, R. T. Spurr determined just how precise and accurate were the assessments that a trained driver could provide. He worked on correlating the numerical outputs of the new brake-testing equipment with a driver's subjective claims of handling qualities. He used the employees of Ferodo to first uncover the accuracy of drivers' assessments of easily quantifiable variables such as deceleration and brake pedal pressure. These were not randomly selected drivers, as Spurr explained: "The testing is carried out by very experienced drivers, some of whom have been testing brakes for ten or fifteen years and all of whom drive many tens of thousands of miles a year. Besides taking these measurements, the drivers also make subjective assessments in their daily reports. These subjective assessments correlate remarkably well with other tests and are therefore very valuable."[31] These drivers could repeatedly decelerate different automobiles with less than 3 percent difference in the rates of deceleration. Spurr's tests, widely cited in the British automotive engineering literature, showed that drivers' assessments of measurable quantities could be very accurate; therefore there was no reason to mistrust the precision of their estimates of more esoteric qualities. The project of matching quantitative and qualitative evaluations became an issue of finding and training drivers and ensuring their safety under the dangerous conditions of road testing. These drivers played an integral role in the development of a common mathematical language to describe vehicle handling.

CONSTRAINTS AND PROBLEMS IN BRAKE TESTING

Engineers working on improving braking systems often faced constraints on their ability to design better systems. Between the 1950s and the 1960s the nature of these constraints changed. The primary obstacle to designing better brakes in the 1950s consisted of limits on engineers' understanding of the way vehicles responded to different braking conditions. Overcoming this shortfall of information required improvements in measuring instruments and led to more sophisticated theories and better mathematical models. These developments changed the nature of designing braking systems. Engineers also overcame their bias against road testing.

Developments in testing technology changed the nature of designing braking systems in general, not only in the area of antiskid devices. First, engineers working on testing instrumentation overcame an industry bias against road testing. Lister's work spearheaded the effort to improve the accuracy and convenience of the road test with his fifth wheel apparatus. His connections to Jaguar, Dunlop, and Lockheed brought his bold approach to road testing into corporate research and development in the early 1960s. As a result of new measuring technology, new designs were produced more quickly and in far greater number in the 1960s. Instead of facing a dearth of information, as they had in the previous decade, engineers in the 1960s had too much information. Their problems lay in separating the useful information from the noise. To produce systems faster they needed instruments that provided the variables their design methods required. Too much interpolation took too much time. For a device to be useful, it had to provide reliable, replicable information in a form readily adaptable to a particular problem. The result of this need for specific measurements led in two different directions: toward more flexible devices which could provide multiple variables for engineers to choose from, or toward a general proliferation of measuring tools which provided more specific types of measurements. The testing devices of the 1960s, best represented by Odier's dynamometer and Segel's and Murphy's variable braking vehicle, were much more generic than the devices Lister had developed. They provided precise measurements of a wide variety of phenomena. The researchers using Odier's or Segel's and Murphy's machines had to decide which conditions to simulate, as they were able to simulate many different braking conditions in Odier's case, and braking systems in Segel's and Murphy's. The number and complexity of machines being designed for measurement increased, while the knowledge created in the process of making these machines also increased along a similar trajectory.

In a review of testing technology presented at the second International Skid Prevention Conference in 1977, Hanns Zoeppritz, an engineer working for the Federal Republic of Germany, made a clear statement of the generic range of testing instruments: "The most important questions always for the choice of measuring equipment or a particular method is the question: What shall we measure? . . . As a consequence of the different goals indicated by the basic question—what shall we measure—different measuring methods are used in road construction and in the automobile and tire industries. This does not exclude using the same equipment."[32] Zoeppritz also reported on the spread of the testing machines to automobile manufacturers, braking component companies, tire producers, and Zoeppritz's own work in designing road surfaces. Zoeppritz documented the exchanges that constantly took place between public agencies and private companies. Between the mid-1950s and 1977 flexible machines designed to measure multiple variables for different types of projects were described in a wide range of journals indicating a wide audience for these technologies.

BRIDGING THE GAP FROM RESEARCH TECHNOLOGY TO ANTILOCK BRAKING SYSTEMS

While much of the historical work on research technology creates a distinction between the production of instruments and the production of new knowledge using these instruments, the case of brake testing technology and ABS blurs this distinction. The machines used to measure the parameters to better understand the dynamics of skidding produced precisely the measurements needed to recognize impending skidding. At its core, ABS is a system for measuring, comparing, and responding. The development of ABS required precise measurements of angular velocity, and ABS itself depended on precise measurements of wheel deceleration. Both testing instruments and ABS had to record changes over time and react quickly to actuate mechanical components. Engineers involved in ABS development recognized that their greatest resource was the technicians and engineers at the testing facilities, and one of the greatest expenses of ABS was setting up tracks on frozen lakes in Scandinavia on which to test equipment. Antilock braking systems evolved in an environment of testing, measurement, and evaluation.

Antilock braking systems had to function without constant maintenance and setup; this was the chief difference between ABS and laboratory testing instruments. Systems had to be as reliable, accurate, and precise as the laboratory and track technology from which it developed, but they also had to be more rugged, maintenance-free, and absolutely failsafe. An instrument failing in the laboratory created far different consequences than an instrument failing on the highway. In developing instruments which measured and compared angular deceleration and torque on four wheels and computers which could nearly instantaneously compare these values, engineers had solved many of the metrology problems inherent in ABS. But packaging these systems of measurement in a small black box in a passenger car was no small challenge. Antilock systems faced very difficult environmental conditions, too. It had to function under wet conditions, in high and low temperatures, and with significant vibration. Depending on its location in the vehicle, ABS also faced being hit by debris. These problems provided plenty of challenges in the development of ABS, even after research-technologists had perfected the measurement systems.

Outside of the problem of producing ABS for a reasonable cost, which proved the greatest challenge in bringing ABS to the market, engineers struggled most often with problems of how to assess performance, since ABS itself constituted a testing system. Testing had to be performed by external instruments, not by the ABS itself, and focused on verifying the responses of the system. Had the system reacted to the conditions in the best and most predictable way possible? This question evolved in three different directions. First, engineers defined performance standards: What was the best reaction for the conditions? Then they focused on two different but complementary ways to test the system against these standards: road testing and computer simulation.

The technologies used for road tests and for generating quantities for the variables used in computer models were quite similar and were both similar to the ABS itself. This convergence brought greater coherence to the community of engineers working on brake testing. Testing machinery proliferated to provide double and triple checks on the responses and accuracy of measurement of the antilock system's performance.

Testing technology and the experience engineers gained in creating instruments and machines for measurement built an important foundation for the development of antilock braking systems. The most common popular presentations of ABS rightly emphasize its contribution to automotive safety, and this creates a tendency to see its development as part of a larger emphasis in the postwar world on building safer automobiles, including the development of seat belts, airbags, collapsible steering columns, and steel-reinforced frames. While safety may have been the motive for developing ABS, none of these other developments in auto safety is related to antilock systems from an engineering or design perspective. No other so-called safety feature is so fundamentally an instrument for measurement and comparison.[33] From the beginning, preventing a vehicle from skidding required a nearly instantaneous measurement of the relative decelerations of the wheels. The knowledge of how to determine wheel speed came from the devices used to test automobile braking systems, tire designs, and road surfaces. In the twenty years between the first International Skid Prevention Conference in 1958 and the introduction of Teldix's electronic ABS in 1978, the development of ABS remained a project with close ties to testing technology. Where one might anticipate that corporate desires would determine the course of research, one finds instead a small, interstitial community of research-technologists both asking and answering the questions that defined the trajectory of ABS's development.

Still, none of the measuring systems documented in this chapter was transformed unproblematically into an antilock braking system, despite the fact that they may have measured the same variables, such as angular velocity, torque, or deceleration, in similar ways. Consumers expect their automobiles to last, with minimal maintenance, for at least ten years or 100,000 miles under harsh conditions, as opposed to the constant assessment and tinkering that a research and development team's instrument might receive. In addition, engineers involved in designing ABS had to create systems that functioned reliably and predictably without the knowledge of the driver. The systems were black boxes in the literal sense of that term, as being noninteractive, input-output devices. The research technology produced by government agencies and universities and private companies was a giant step closer to ABS than an idea in an executive's mind, but not yet a complete transformation.

five

From Things Back to Ideas

Constructing Theories of Vehicle Dynamics

In 1930 A. W. Frehse wrote an article for the Society of Automotive Engineers in which he placed his hopes for the future in the mathematical prediction of braking system performance. He wrote, "Accurate mathematical expressions and formulae now supplant the old cut and try method. . . . Armed with these mathematical tools, a designer can proceed with his designs without the usual misgivings and uncertainties that characterized former brake design."[1] Thirty years later G. Jones of British Motor Car Corporation wrote an article for the *Proceedings of the Institution of Mechanical Engineers* showing little advancement in the mathematical approach to brake design.[2] This is not to say that Jones's approach was not technical or sophisticated. On the contrary, Jones's article was highly theoretical, perhaps even explanatory in the scientific sense, just not particularly mathematical. The article was widely criticized for its "totally descriptive" approach, with a number of engineers writing to plead with Jones to follow up his article with one that would marry the experimental and descriptive approaches he used with mathematical and analytical models that automotive engineers were borrowing from aeronautics.[3] In the thirty years between Frehse's and Jones's articles, the pressure to mathematize had grown. The community has responded to this pressure; since 1960 few articles have been as devoid of mathematical models as Jones's. His article was the last of a breed, published even as computational methods of describing the behavior of a vehicle were taking over his field. Jones never again published in the *Proceedings*, though he continued working as an engineer in the automotive industry. This chapter looks at the way the ABS knowledge community defined and shaped the mathematization of automobile behavior and the effects these new theoretical questions had on both the ABS community and the devices they were designing in the 1960s.

The principal pioneers of the mathematical modeling of vehicle dynamics were William F. Milliken and David A. Whitcomb of the Cornell Aeronautical Laboratory (CAL) in Buffalo, New York. In 1955 Milliken, Whitcomb, and four other colleagues from CAL were invited to present their work at Institution of Mechanical Engineers meetings in several cities throughout England. A year later new versions of their lectures were published in the *Proceedings of the Institution of Mechanical Engineers*, along with lengthy transcripts of the discussions that followed and written comments to CAL and responses from Milliken and his colleagues. At CAL Milliken headed the full-scale division; in contrast to the scale model division, which was occupied largely with wind tunnel simulations and the calculations necessary to overcome scale effects,[4] the full-scale division was essentially a testing division, with two departments, flight research and vehicle dynamics. Whitcomb worked in Milliken's full-scale division as the assistant department head of vehicle dynamics. Together Milliken and Whitcomb began a trend to apply and modify theories of handling and stability from aeronautical research to automobile design.

When Milliken and Whitcomb first began applying aircraft stability modeling to autos, they were surprised at the lack of sophisticated theoretical tools used by automobile designers. They wrote, "At the present time, after 70 years of automotive engineering, there is still little universal agreement in the definition of specific handling qualities, or what constitutes desirable handling, and on the contribution of design to handling. In some questions, apathy prevails toward the complexity of analytical procedures that might be utilized for the study of handling, whereas in others the belief exists that knowledge currently available in practical design is adequate."[5] Clearly neither Milliken nor Whitcomb found their methods overly complex, nor did they find existing tools adequate, so they set out to invent what they referred to as "a formal science of automobile handling."[6] Their work defined the field of vehicle dynamics.

In their original series of articles in the *Proceedings* in 1956–57, Milli-ken and Whitcomb not only laid out their credentials as aeronautical researchers, but also presented a plan to apply theories of aircraft sta-bility to models of automobile handling.[7] In their articles, subtitled "A General Introduction to a Programme of Dynamic Research," they wanted to set an agenda. In addition to championing the application of their own work to auto design, they encouraged both automotive and aeronautical engineers to develop new dynamical theories. They wrote, "[The science of automobile handling] is an engineering activity largely empirical, residing within the experience of the designers and the test drivers and the pages of a small literature of technical papers. Finally, it is a field of unquestioned fascination, replete with unknowns and opin-ions, all of which revolve around the question, 'Why does the auto-mobile behave as it does?' "[8]

Milliken and Whitcomb presented their long-range program as seven discrete stages. Most of these stages focused on improving the predictive accuracy of mathematical models. Because their work was oriented to-ward prediction, they also put out a call for better data and more sophis-ticated measuring instruments. Their presentations acknowledged that the data provided by the RRL was the most accurate available, and they urged the RRL as well as MIRA to continue funding and providing data.[9] Milliken and Whitcomb were explicit in reminding their readers that theoretical tools did not substitute for empirical data; they relied on it. They also impressed upon their readers the importance of equating their mathematical models with physical systems, writing in 1956, "Implicit in their [Milliken's and Whitcomb's] handling theory is an explanation of the physical nature of the motion of the vehicle."[10] Milliken reiterated this point in his response to comments on his paper, presenting a heavily illustrated version of his theory in the discussion section of the *Proceed-ings*. Explaining his reasons for representing the theory, he wrote:

As brought out in the informal discussions there had been a great deal of curiosity among practising engineers about the physical nature of automobile behaviour. To the degree possible, most engineers desired a physical as well as a mathematical insight into car motions. . . . It was possible to go still further in the visualization of the interactions of the various physical factors that con-tribute to car motion, and the following comments and illustrations have been prepared with that objective in mind.[11]

Milliken worked hard to make his approach palatable to skeptical but practical old-timers.

Interestingly, Milliken and Whitcomb based their science of vehicle dynamics on a historical model. They claimed that developers of vehicle dynamics should attempt to replicate the path used in the design of aircraft, surmising that theories of vehicle dynamics in 1955 were at the stage aircraft stability studies had been in the 1920s.[12] Using work in aircraft stability as a model of how to build vehicle dynamics, they thought automobile theories would catch up quickly. However, this attempt to equate aircraft and automobile design produced resistance to CAL's methods, as many automobile engineers were not willing to make this leap. At least a dozen comments in the *Proceedings* attack this claim precisely. For example, "[Milliken and Whitcomb] had also stated that they considered progress in parallel matters in other technical fields (presumably in the aeronautical world in particular) had been much greater than in automobile engineering, a comment which seemed to require substantiation by them as what constituted a parallel condition could be very debatable."[13]

Throughout their papers and especially the discussions, Milliken strove to convince engineers that mathematical models were *useful* design tools. The considerable effort he put into repeatedly making this claim gives a clue to the amount of resistance from automobile designers he and Whitcomb anticipated. The comments in the discussion sections of the *Proceedings* provide further proof that not all engineers welcomed such a mathematical approach, especially one borrowed from a different industry with its own distinct engineering communities.[14] Resistance to mathematization took several different forms. One concrete criticism of the CAL approach was that CAL researchers performed all their calculations on an IBM computer and had programmers to process empirical data,[15] which was essential to their work on wind tunnel calculations. Plugging in algorithms from automotive testing posed no programming difficulties for CAL. But for engineers at British automobile companies, using a computer often posed an insurmountable problem; their companies did not have digital computing facilities and had no plans to acquire them, let alone having a department of programmers to aid them. Milliken dismissed this objection, claiming that the use of the computer on automotive problems was the equivalent "of driving tacks with a sledge hammer."[16] Others complained that automotive design budgets were minuscule next to aircraft design budgets, largely because automotive prototypes were easily and inexpensively constructed, unlike aircraft and ship prototypes. Auto engineers questioned the need to construct mathematical models when physical models were easily produced. Milliken responded simply that paper designs were the wave of the future; he was right.

Leonard Segel was another member of the CAL group invited to present his work in England. Segel, the co-inventor with Ray Murphy of the variable braking vehicle at the Highway Safety Research Institute, earlier in his career was head of the vehicle dynamics department at CAL. His paper to the Institution of Mechanical Engineers focused on the correlation of theory to the experimental results of steering response. Like Milliken and Whitcomb, Segel arrived in England as a staunch defender of mathematical modeling. He claimed that practicing engineers were often too close to problems to understand how to simplify them for mathematical modeling. Segel wrote, "It takes a certain amount of engineering courage to make a problem susceptible to analysis."[17] Experienced engineers accustomed to empirical methods assumed that the simplification or idealization necessary for mathematical modeling was a real weakness. Segel's somewhat superior and patronizing tone did not convince them otherwise. However, his central argument, if not his style, tried to make engineers more comfortable with simplifications for mathematical modeling by showing close correlations between theoretically and experimentally derived results.

Segel's paper was less programmatic than Milliken's and Whitcomb's and focused equally on testing and theory. In a paper published in the *Proceedings*, he presented his view of the uncomplicated nature of testing: take a car, fill it with instruments, make measurements, produce data, and prove the mathematical model, which was constructed a priori.[18] Segel wanted to reduce, if not eliminate, road testing from the engineer's job description. He claimed his goals were "the development of a mathematical model to replace the physical automobile and the utilization of full-scale response measurements to check the dynamic behaviour predicted by the derived equations."[19] Full-scale response tests were the critical ingredient to support Segel's theories. Once the predictive powers of a theoretical tool were proven, there was little need to test the theory continually. Measurement was important largely as an ingredient.

Segel's presentation to the Institute of Mechanical Engineers in London in 1955 focused only on steering; he would address braking a decade later at the Highway Safety Research Institute. But his experience in building a mathematical model from empirical data produced a model for how to simplify a complicated real-world problem. First, he assumed the auto behaved linearly. Strictly speaking, this was not the case, yet Segel showed that assuming linearity for moderate lateral motion yielded results quite close to those he had measured. Still, beyond his statements critical of trial-and-error engineers, his assumption of linearity was his most vulnerable point. A decade later, models would assume nonlinearity as engineers gained access to the computers that made nonlinearity and partial differential equations tractable. In addition, as models became more sophisticated, initial idealizations were constantly questioned, and the assumption that moderate lateral motion could be idealized as linear behavior did not pass scrutiny.

Despite its drawbacks, Segel's presumption of linearity held benefits for modeling the behavior of the car. Linearity allowed him to use superposition to break stability problems into smaller parts and add the results of each analysis to calculate the behavior of the whole. His breakthrough was in being able to break the dynamic system down into parts small enough to yield results comparable to the instrument readings his test vehicles provided. Even though his approach was not overarching like Milliken's and Whitcomb's, his work showed engineers what to do and how to go about designing new models. In this sense his model was more useful than the program for research presented by Milliken, which told engineers only *what* to do and focused far less on *how* to do it. The work of CAL researchers set the tone for investigations into vehicle dynamics, but none of the initial work addressed the problems of modeling braking.

NONLINEAR BEHAVIOR AND THE INCORPORATION OF BRAKING INTO MODELS

In 1960 K. N. Chandler of the Road Research Laboratory published an article titled "Theoretical Studies in Braking." Taking up Milliken's and Whitcomb's challenge to produce a mathematical model of a simplified vehicle, Chandler's model tried to predict stopping distance. Drawing from considerable data produced on stopping distances at the RRL in the 1950s, Chandler had data of the stopping distances that had been produced on different road surfaces at different speeds. The difficulty he faced was to produce a solvable mathematical model that agreed with known results. His model of a vehicle with locked wheels predicted stopping distances relatively accurately, but when he modeled the Maxaret antiskid device his calculations yielded considerably longer stopping distances. Why? Chandler had several explanations.

Unlike researchers working on mathematical models at CAL, Chandler worked without a computer. In dealing with partial differential equations and nonlinearity, he had to make approximations and simplifications to solve the model. He believed that some of the error was produced this way. He also questioned the model itself.[20] In the end he analyzed four different reasons for the predictive failure of his model. First, he could not take into account the automobile's suspension. The suspension continually changed the loads on each wheel, making it impossible to calculate when a wheel was going to lock up. Second, the model assumed that all the wheels worked together, responding to the same conditions. Again, because of the suspension, they did not. Third, unconstrained by available valve technology, Chandler's model of the Maxaret could pulse much more quickly than the real system. While this was desirable, it meant that Chandler's model did not correspond to the physical system. Chandler's model indicated that faster cycles of braking would be beneficial, but modeling the Maxaret inaccurately ruined the chance for correlation between testing and theory. Fourth, his model did not account for the self-energizing character of the braking system.[21]

Chandler's model first defined wheel slip as the difference of the quantity of 1 minus the angular velocity of the wheel on the road divided by the quantity of angular velocity of the wheel rolling freely.[22] These two variables of actual rolling and idealized free rolling had to be produced separately in different tests. During braking, slip had to have a value between 1 and 0; it was 1 when locked up completely and 0 when rolling freely. Chandler then created four equations that defined the motion of the wheel during braking. These equations could be integrated to produce a value for the torque of the brake shoes on the wheel. Stopping distance varied directly with this variable. However, Chandler's model was a steady-state model, assuming that speeds did not affect the friction between road and tire. While theoretically friction does not change with velocity, it does change as rubber tires heat up, which is dependent on velocity among other factors. Chandler's calculations also assumed that the brakes were either fully applied instantaneously or completely off. In short, his calculations did not take into account the complexity and ever-changing nature of the braking system. In fact, Chandler was not trying to model the braking system, he was merely trying to create a set of calculations which would predict the stopping distance of the vehicle.

One wonders why Chandler published an article that he admitted failed to predict the variable he was seeking. One reason may have been to lend support to the RRL's interest in the Maxaret. His article did do a much better job explaining the operation of the Maxaret than Lister's ever did. But Chandler's work cannot be seen solely in this light. Chandler wanted to show the possibilities of modeling the braking system; other engineers could build from his mistakes. Thus Chandler showed a certain amount of courage in presenting a model he admitted was flawed.

J. R. Ellis, director of automotive engineering at the Royal School of Aeronautics in Cranfield, England, made the most important attempt to model the behavior of a braking system more accurately. Ellis had been a vocal participant in discussions of vehicle dynamics from the start. He championed the construction of mathematical models for vehicle dynamics, urging automotive engineers to strive for "a comprehensive picture of vehicle stability and control response."[23] Like the engineers at CAL he worked in an institution where the marriage of aeronautical theory with automotive empirical data was not only possible, but natural. Like Milliken and Whitcomb, he was convinced that the empirical and theoretical had to grow together. Unlike engineers who wanted to stick to completely empirical, cut-and-try engineering, Ellis argued for more complicated mathematical models while emphasizing the need for continued improvements in measurement technology. Working with Robin S. Sharp, a recent Master's degree student, Ellis and Sharp wrote, "In the teaching of vehicle dynamics it is desirable that a series of practical experiments and demonstrations should be made in order to demonstrate the validity and application of the theoretical analyses."[24] However, Ellis was critical of Milliken and Whitcomb in 1955, accusing them of oversimplification. He questioned Milliken's and Whitcomb's assumption that their vehicle was a steady-state system, urging them to accommodate the fact that weight distribution and forces change with velocity. He argued that the motion of a vehicle could never achieve a steady state. When he then presented his own model in the *Proceedings* in 1963, he was answering the challenge issued by Milliken and Whitcomb to produce new theoretical tools.

Ellis was trying to build a comprehensive mathematical model of the automobile's motion. His model would accommodate braking, steering, and acceleration, while the earlier models of the CAL engineers dealt only with steering and lateral motion. In the first sentence of the abstract of his article, Ellis wrote, "When the performance of a vehicle which may be braked, accelerated and steered is considered it is necessary to use large displacement theory to cover all phases of the motion, and the restrictions implied by the constant linear analysis of Milliken *et al.*, are not acceptable."[25] Later in the article he refined his criticism of the CAL work, writing, "A failing of the Cornell theory is that in the present state it is unable to differentiate between changes in the distribution of vehicle rolling stiffness between the axles, since a total roll stiffness is used in the calculations."[26] The problem with using a single value for roll stiffness was that the calculations then failed to predict the rotation of the car when the rear brakes locked up.

Ellis used an ENIAC 11 analog computer to model the hypothetic passenger car he described, but claimed that the computer was really unnecessary and could be replaced by tabular calculations. This may have been true, but few automotive engineers were equipped to deal with partial differential equations of nonlinear deflections. However, since Ellis concentrated on studying braking, his work brought vehicle dynamics modeling into the community of engineers working on braking. To these engineers, Ellis's tools represented the first attempt to model mathematically what had been shown to happen empirically. His method predicted the rotation of the vehicle when its rear wheels locked. It also predicted the loss of directional control when the front brakes locked. Ellis then emphasized the promise of skid prevention devices by building a model of a system that applied and released the brakes repeatedly in a short period of time. His model indicated that the driver would not lose control of the car under these circumstances.

Both Chandler's and Ellis's models explicitly tried to show the advantage of preventing brakes from locking. As was known empirically, eliminating brake lockup had three advantages: it shortened stopping distances, retained stability, and allowed the driver to retain directional control of the car, or steerability.[27] By the mid-1960s most attempts to model the braking behavior of an auto used these three factors as their litmus test. Chandler's model predicted none of the factors. Ellis's model did not try to show improvement in stopping distance but did predict the retention of directional control.

DEVELOPMENTS ON THE EUROPEAN CONTINENT

Jean Odier, employed by Ferodo in France, published several attempts at modeling stopping distance and steerability in the 1960s. Unlike Chandler and Ellis, who made explicit efforts to show the advantages of antiskid mechanisms, Odier focused on showing the reasons for the stability and instability of the whole braking system. In the early 1960s he was critical of automotive engineers for being reluctant to use theoretical tools to study anything other than braking efficiency. According to Odier, the promising results yielded by mathematical studies of efficiency should have spurred engineers to broaden their mathematical approaches to other braking issues.[28] Like other engineers focusing on dynamics in the 1960s, Odier was an ambassador for mathematical modeling. Like Milliken, Chandler, and Ellis, he expressed his dismay at the slow speed with which automotive engineers took up mathematical and theoretical models.

As someone who was moving between theoretical and empirical studies of braking, Odier was most interested in describing and explaining peculiar aspects of braking systems. For example, he asked why a car that was stable at a moderate speed became highly unstable at high speeds.[29] He thought that many of the defects of a braking system were exposed only when certain conditions were present in combination. He wrote, "The relationship between [friction coefficient, coefficient of brake instability, and the reversibility characteristics] determines the ratios that have to be complied with between various parameters to avoid defects."[30] Odier published and presented papers in French, English, and German and moved easily between the vehicle dynamics and metrology communities. His work was widely cited as a model of the utility of vehicle dynamics. Moreover, the types of problems he chose drew enormous attention, and consequently he raised interest in predicting the counterintuitive.

It is interesting to note that up to this point the center of the vehicle dynamics community lay in England. Even engineers based outside

England, such as Odier and those at CAL, published their earliest findings in the publications of the Institute of Mechanical Engineers, the British professional society. Articles appeared in American publications but were often derivative of what had appeared in England months or years earlier. There were other locations for publication, principally Germany, although German work on mathematical models did not begin to appear in English publications until the late 1960s. In several ways the German models were more sophisticated, but the way they modeled the behavior of the automobile harkened back to Milliken's and Whitcomb's call for a model that represented the whole handling problem, a much more comprehensive approach. And while German work was not often published in English in the 1950s and 1960s, German articles commonly cited the work of Milliken and Whitcomb; the avenue of transmission of models and ideas appeared to be one-way for about a decade. Nevertheless, engineers working on vehicle dynamics in Germany built closer ties to the antilock community, so their work ultimately had a greater influence on the design of antilock systems.[31]

Working in the 1960s in Germany, Manfred Mitschke tried to construct comprehensive models of the total dynamic behavior of a vehicle. He wanted to realize the idea of Milliken and Whitcomb, by making one very complicated model, which would accommodate braking, handling, and stability. Unlike Ellis and Chandler, whose work focused exclusively on braking, Mitschke's model tried to relate steering and braking together in a single mathematical model. Mitschke had been an engineer at Bosch from 1960 to 1966 and simultaneously held positions as Privatdozent at the Technische Hochschule in Braunschweig and Karlsruhe. In 1966 he left Bosch to oversee the Institut für Fahrzeugtechnik at TH-Braunschweig. Under his leadership in the late 1960s, Braunschweig emerged as one of the primary centers for research into vehicle dynamics. As a research leader as well as a director of a research facility, Mitschke had a position of high visibility in the development of vehicle dynamics models that accommodated both braking and issues of directional stability. His primary goal was to model directional stability, taking into account both the action of a driver in steering the vehicle and the characteristics of the braking system and road that led to directional instability. Mitschke asked whether it was possible to model an "ideal driver" trying to countersteer a sliding vehicle to overcome the loss of control. Could this common occurrence be reduced to a set of equations? In an extensive review of the literature on directional control in the *Automobiltechnische Zeitshrift* in 1968, Mitschke tried to correlate the existent theoretical and empirical literature and outline where the inconsistencies were.[32] For Mitschke, disagreements between computational results and empirical testing were the preferred areas for immediate work in the field of vehicle dynamics. While other vehicle dynamicists would have agreed, few put it in such stark terms.

Mitschke broke the problem of directional stability into several parts: idealizing the behavior and responses of the driver, examining the statics of the vehicle, and investigating the dynamics of directional control on the road surface and tires in each relevant component system and finally in the vehicle's comprehensive behavior. With this approach he could detail where linear approaches would work and where the model had to be nonlinear to represent the critical interactions. Mitschke's project seemed to be broken up in ways remarkably similar to that of the RRL a decade earlier, though he never cited that work in any revealing way. In addition to reviewing the literature and pointing to the direction for further investigations, Mitschke published his own models of directional stability in 1968, where he concentrated on the interaction of the driver with the lateral forces created by braking while cornering.[33] His interest in modeling the driver sets his work apart from the other dynamicists of the 1960s. Mitschke saw driving as a cyclical process—a feedback cycle, if you will—and believed that eliminating the driver's responses was one important cause of poor correlation between computational and test-track findings.

In 1970 Mitschke published an article of immense importance to the development of antilock systems, "Blockiervorgang eines Gebremsten Rades." Here he took an approach different from his earlier work in that he was no longer trying to construct a comprehensive model of handling. Instead, this model tried to accommodate all the factors of the dynamics of a wheel, focusing on depth rather than breadth of examination. Mitschke's model dealt with braking, cornering, acceleration and deceleration, and the interaction of road and tire and took a nonlinear approach to lateral forces.[34] By far the most complicated approach to modeling the wheel, Mitschke's model accounted for many of the factors that had been disregarded in the past in the name of idealization.

Mitschke's model did not appear out of nowhere. This is evident in the way he addressed many of the problems earlier dynamicists had brought up in their work. Mitschke took previous models and addressed their shortcomings; then he designed new ways to approach these problems and incorporated new, more sophisticated, mathematical constructs into his model.[35] The way he designed the model of the "braked wheel" shows that mathematical models were *designed*. Like the engineers working on the design of things, Mitschke made choices about what he had the resources to represent, incorporating effective parts of older models. The engineering practices he used in designing his model of the "lock-up process of the braked wheel" were similar to those used by instrument makers in producing gradual improvements to their machines. Mitschke consolidated older work and added new complexity to his model, and therefore made it more useful to engineers working on antilock systems. His model of the braked wheel was an innovation in the construction of mathematical models. While there was some continuity with older approaches, Mitschke took vehicle dynamics in a different direction. This redirection proved interesting to a new audience of antilock braking designers, who were seeking applicable theories in 1970.[36]

By the end of the decade of the 1960s vehicle dynamics had evolved to the point where it was being taught as part of the curriculum in automotive engineering, which was usually a specialized track in mechanical engineering. It did retain an academic bent for longer than its originators wanted. Starting with Milliken's and Whitcomb's work, vehicle dynamics had gone from a field with little appreciation of the problem of braking to a field consumed by the problem of braking in the late 1960s. While there were engineers working on the dynamics of steering and handling, research on braking was far more prestigious and was a common theme for plenary sessions at conferences devoted to the field. In addition, vehicle dynamics, as it related to braking, often grabbed the spotlight at the increasing number of conferences on automobile safety. Ralph Nader, in his famous indictment of the American automotive industry in 1965, *Unsafe at Any Speed*, mistakenly claimed that theoretical studies proved the efficacy of antiskid systems.[37] Actually, vehicle dynamicists thought that the continued development of antilock systems validated their work and showed the utility of vehicle dynamics. The resistance theoreticians had faced until the mid-1960s was completely gone, and in its place was an enthusiastic optimism about the promise of mathematical models. Journals in the 1960s were deluged with submissions of mathematical models of every aspect of vehicle motion. By 1970 separate journals had been formed which were devoted exclusively to the field. What happened in the mid-1960s to explain the relatively sudden acceptance of mathematical models of car behavior?

By and large, an answer can be found in the increasing availability of the digital computer, coupled with the diffusion of FORTRAN.[38] Earlier, engineers at CAL had been criticized for using digital computing, which most automotive engineers did not have access to. By the mid-1960s even companies that could not afford a large computer could purchase shared computer time. In a shift that can be seen throughout engineering disciplines, the accessibility of the computer in the 1960s meant that engineers began to seek out problems solvable by computers instead of avoiding mathematical models requiring lengthy iterations. For most engineers, two kinds of equations went from deadly to desirable: partial differential equations and nonlinear equations. While these types of equations were practically impossible to solve by hand, computers could iteratively approximate solutions quickly. Inexpensive and common access to computers encouraged modelers to seek solutions that intentionally included these forms. For the engineers involved in constructing vehicle dynamics, this opened up a number of ways to avoid the simplifications that undermined Chandler's models.

THE TECHNISCHE HOGESCHOOL AT DELFT: A LABORATORY FOR VEHICLE DYNAMICS

Like Jean Odier, who asked *why* questions about the behavior of a braked vehicle, W. T. Koiter and H. B. Pacejka of the Technische Hogeschool in Delft went back to early studies of braking to determine what was still unexplained. Koiter and Pacejka were faculty at the Laboratory of Engineering Mechanics at the Technische Hogeschool beginning in the mid-1950s. They went all the way back to Bradley's and Wood's work after World War I at the original RRL, which asked why a car rotated when its rear wheels locked.[39] They knew that the fundamental reason for this was the lower value for sliding friction of the rear wheels compared to the continued rolling friction of the front. But Bradley and Wood had never addressed the directional attitude of the car. Why would it rotate in one direction and not the other? Like Mitschke, Koiter and Pacejka wanted to build stability and handling theories into braking dynamics, and like Mitschke's models, their models included lateral forces created by cornering or by the driver's response. This made for a very complicated model, especially as the deviations became large. Unlike Mitschke, who preferred to make models linear whenever possible, Koiter and Pacejka designed models that were unabashedly nonlinear, and once the lateral displacements became large at high speeds they required a computer for solution. Fortunately by the late 1960s this was less of a drawback than it had been a decade earlier. Thus Koiter's and Pacejka's models were able to accomplish something very important; they predicted nonlinear automobile behavior. Following Odier, mathematical models were increasingly sophisticated and oriented toward providing an understanding of the more peculiar and unpredictable behavior of a car. This function was much more useful than the theories that merely provided correlation with known empirical results. Engineers at Delft took a new approach in the 1960s and developed a series of models even more important to antilock research than Mitschke's.

In the mid-1960s Ramachandra Rao Guntur and his colleagues revisited Chandler's model. Guntur was the chief scientific officer at the newly formed Vehicle Research Laboratory at the Technische Hogeschool in Delft. The Vehicle Research Laboratory had evolved from the Laboratory of Engineering Mechanics since so much of the work there had focused on vehicle dynamics. Guntur collaborated with H. Ouwerkerk and focused on designing a mathematical model that could be used to predict the impending or imminent skidding of a wheel in a theoretical setting.[40] However, creating a theoretical tool was not Guntur's only goal. He figured that if he could construct an algorithm to predict skid on a computer in the Vehicle Research Laboratory, he could use that same algorithm to predict imminent skid in a moving vehicle. Experimenters were already putting computers into cars for measurement. Why not also use the computer to process signals and prevent skidding altogether? Guntur was marrying vehicle dynamics to a control system for skid prevention. Now vehicle dynamics, which played a crucial role in creating new knowledge about the mechanics of skidding, was playing a role in the design of skid prevention systems.

Guntur published numerous internal reports on the connections between models of vehicle dynamics and control algorithms, but he focused his publications outside the Technische Hogeschool on the nature of adaptive control logic.[41] He and several other researchers were working on adaptive control systems for antilock braking. An adaptive control system can adjust its parameters depending on the conditions of the moment; instead of having a preset torque or deceleration threshold at which point the brakes begin pulsing, adaptive control systems consider variables such as the speed of the car or the distribution of weight and braking ratio between the front and rear wheels. Guntur argued that adaptive control was necessary, otherwise a braking system operated at peak efficiency only in a very narrow range of conditions.[42] If vehicle dynamics could predict skidding under a variety of different road conditions, and Guntur could incorporate this predictive function into a control system, then vehicle dynamics would more than fulfill the promise that Milliken and Whitcomb saw in it.

Several antiskid researchers were grappling with the problem of how to *predict* skidding. It is easy to conceive of how to predict skidding by figuring out how much a tire is slipping on the road as a percentage. For example, if a tire is slipping 100 percent, clearly the brake has locked and the driver has lost control of the vehicle. Engineers predicted that if a wheel was slipping less than 50 percent against the road the driver could maintain directional control; if a wheel was not slipping at all, braking efficiency would be reduced.[43] Consequently, an optimal amount of slip had to be calculated and generated by the braking system. The problem with this method was that wheel slip could not be easily measured; it existed largely as a theoretical construct. Guntur asked which currently measurable parameters correlated with wheel slip. His mathematical model took inputs from sensors on the auto in the form of angular and linear velocity, weight distribution, torque, and rates of brake application to calculate an approximation of wheel slip and compare it to a calculated value for critical slip, which would itself vary with speed and weight distribution. In addition, the model reflected the unique geometry of a particular model of automobile. Guntur called the initial calculation and comparison of wheel slip "the predictive mode." If calculated wheel slip was less than critical slip minus some safety factor, then the brakes would operate normally. However, if slip started to exceed its limit, then the control of the braking system would be given over to a device that pulsed the brakes rapidly in order to decrease the amount of slip. The pulsing continued until the computer recalculated the slip at an acceptable level. The computer constantly sampled the variables and performed the slip calculation until slip dipped back under the limit, which was also being constantly recalculated. Guntur called this process of continually sampling the motion variable of the car "reselection."[44]

Guntur was able to transform his calculations of vehicle dynamics into control algorithms because he possessed several advantages over earlier theorists. He and his colleagues at Delft used existent antiskid devices in their work modeling vehicle dynamics. In an article in a journal devoted to vehicle dynamics, Guntur presented a list of antilock systems under development and the status of their control systems.[45] None could function without some kind of electronic control, although most of the systems were analog systems. Guntur wrote, "A modern anti-skid system consists of a sensor, a control unit, and a brake pressure modulator."[46] After nearly two decades of development, vehicle dynamics had become a critical research area in the development of antilock braking systems. Furthermore, the machinery Guntur was using to test his algorithms operated in a fashion very similar to antilock systems then in commercial development.

By 1970 engineers were coming to Guntur and his colleagues to discuss how vehicle dynamics could inform their designs for antilock devices, especially in the area of adaptive control.[47] Many engineers came to Delft to test their systems. Those in the commercial sector in particular had a great deal to gain from the concentration of engineers working in vehicle dynamics at Delft, and the give and take of knowledge at a university lab seemed a reasonable price. Offering use of the lab at Delft to commercial projects meant the researchers at Delft were at the forefront of commercial R&D, and the engineers who came in from corporations were privy to the newest theoretical tools. As had been the dominant communal mode in ABS work, public and private sector organizations worked together without incident.

In 1955 researchers working in the nascent field of vehicle dynamics were just beginning to hear about mathematical models being generated by research into aerodynamics and the handling characteristics of airplanes. While it is important that the first researchers to apply aeronautical theories came from environments where aircraft research was being undertaken, the development of vehicle dynamics quickly split from these early connections to aeronautical research. What developed as vehicle dynamics was more akin to the work that the RRL had aimed to do: creating predictive models of the interaction of the road, the driver, and components of the car. Incorporating the complicated and often unpredictable reactions of a driver into mathematical models of the car made these models more accurate. But more important, the increasing sophistication of mathematical models made them particularly useful to researchers interested in predicting the stopping distance and imminent skidding of a vehicle. Interest in vehicle dynamics arrived at its apex just as commercial antilock systems started to appear on the market in the 1970s. Consequently, the crossover between building mathematical models and designing antilock systems played an important role in making functional systems, and perhaps even more critically in mapping out the antiskid community, by defining one area of common, non-proprietary knowledge.

six

Learning from Failure

Antilock Systems Emerge in the United States

A common issue that arises in the history of science and technology is the priority dispute.[1] In the least interesting and most easily solved cases, these arguments relate to who filed which patent on which date. Still others focus on the nature of the relationship between similar artifacts whose development seems completely independent; the best examples are disputes over dyes in the nineteenth century, the invention of the internal combustion engine, and most famously Newton's and Leibniz's simultaneous development of two systems of calculus. The invention of commercially available antiskid devices is characterized by a series of priority disputes of yet another type. These debates are perhaps more interesting to the historian, since they hinge on issues of definition. Claiming one had developed the first antiskid device depended on one's definition of what constituted an antiskid device. Because different engineers and different companies defined the problem of skidding and its possible solutions differently, there were multiple valid claims as to who produced the first commercial antiskid device. There was also a debate about what *commercial* meant: Did it simply mean that the device was available on the market without regard to cost or how available the system was, or did it mean that the device was widely available, that is, commodified? Given that nearly a dozen different antiskid devices were introduced between the mid-1960s and the mid-1970s, looking at competing claims of which was the first provides a way to examine the practitioners' collective construction of a definition of an antilock system.

By the 1960s there were no fewer than a dozen different antiskid devices under commercial development in Europe and the United States. The majority of these machines were eventually introduced to the market in the 1970s. Even before they were commercially available, nearly all claimed to represent the advent of a new era in braking design and emphasized the radical newness of their designs. Looking at the advertisements and articles that accompanied the introduction of any number of systems, one would have difficulty discerning that several companies were engaged in developing antilock devices simultaneously; each was presented as new, novel, and one of a kind. For the antiskid market in the 1970s, being first seemed to hold great promise, and companies tried to claim priority whether or not they could justify it. Chrysler, Ford, Bosch, and Teves all claimed to be making *the* breakthrough. This shows the importance of priority in innovation, and also perpetuates the impression manufacturers want to make that there is a single keystone to introducing a new product. In the case of ABS, it is quite difficult to justify any one development as *the* breakthrough, even though engineers and marketing executives behaved as if there was only one. Every company confidently tried to introduce the first *true* antilock device, or better yet, a *true system* to prevent skidding. While these claims must be taken with a grain of salt since they are more marketing claims than engineering claims, it is interesting to examine how engineers did justify their own design as the first. Obviously, any machine could be claimed as "the first" if the definition of that machine were peculiar enough.

It is important to remember Bucciarelli's statement that how a thing works is inseparable from what it is. If all the antilock systems introduced during the 1970s worked differently they could all make claims to be the first of a new generation of braking designs. These differences, some greater than others, were the basis of a series of priority disputes centered on the question of who pioneered antilock systems.[2] To a large extent the question of who was first was more than a simple priority dispute, because the controversy rested not on which design was chronologically first, but on which design was first *and* constituted a true skid prevention system. In the end, the claim of priority was settled by the Sperry Award of 1993, given by the SAE to engineers at Teldix and Bosch in Germany. However, it is important not to see the Sperry Award as a completely external validation. In fact, the SAE is a large professional organization that encompasses the ABS knowledge community;

its journal and *Automobiltechnische Zeitschrift* were the two most common venues for publication of ABS research. Seen in this way, the Sperry Award publicly marked an achievement that engineers inside the ABS knowledge community accepted. The award shows that the ABS knowledge community of the 1970s was mature enough to adjudicate claims of expertise by practitioners within the community.

DEFINING ABS

The first devices aimed exclusively at reducing skidding were developed by individuals whose work has already been discussed. By most accounts, the Dunlop Maxaret was the first purpose-built antiskid device, introduced commercially in 1966 on the Rolls-Royce Silver Shadow. The Maxaret then became available on selected Rolls-Royces and Jaguars and was probably most notable for its inclusion on what some have called the most innovative automobile of the twentieth century, the Jensen FF.[3] The Jensen was also one of the first commercially available four-wheel drive vehicles, but it was a hand-built car, and only 330 were ever produced. Although the Maxaret had been available on several different makes and models of cars, fewer than a hundred were ever sold. Due in part to the small number of units produced, many engineers saw the Maxaret as a prototype rather than a production antiskid device. Along these lines, a Rolls-Royce, Jensen, or even Jaguar was not going to have any impact at all on traffic safety, since their design, manufacture, and cost shared more design and production characteristics with racing cars than with mass-marketed automobiles. Consequently, later engineers claimed that theirs were the first *production line* antiskid devices to be introduced to a broad market. The effort to produce an affordable, mass-production antiskid device has always played an important part in setting design criteria for antilock systems since engineers recognized that improvements in safety would come only with widespread use of antilock systems. If antiskid devices were in use on fewer than 10 percent of vehicles, they simply would not make a statistical improvement in skid prevention.[4] So making skid prevention available to a wide market was more than a marketing dream; it was part of what constituted good system design.

Complicating the issue further were claims in the 1960s that devices with different design objectives than skid prevention could be considered antiskid devices if they had the potential to reduce skidding under certain conditions. Brake-limiting and -proportioning devices have been given credit by their designers for combating skidding, although clearly they do not modulate an automobile's brakes in order to prevent skidding. In these devices, skid prevention is an ancillary benefit from a device designed to prevent overbraking regardless of road conditions. These machines do not detect imminent skidding or do anything to truly prevent skidding.

Hans Strien, an employee of Alfred Teves, GmbH, one of the main producers of antilock braking systems in the 1980s, was involved in both the improvement of braking systems in the 1950s and 1960s with the diffusion of disc brakes and the introduction of brake power-limiting and -reducing apparatus. In 1961 he presented a paper at the SAE meeting in Detroit that made it clear that antiskid devices and brake-proportioning devices should be differentiated:

These [power-limiting and -reducing] devices are frequently offered as "antiskid devices" or even as "locking-prevention devices." Since it is by no way clear as to whether the customer is purposely deceived for business reasons, or whether engineers actually have different conceptions about the function of an anti-skid device, it would certainly be a useful task for the SAE and this Congress to establish clearly once more terms and definitions of the anti-skid control, even if this should have been done at some other place already.[5]

The SAE never took up Strien's advice to establish a definition of antiskid control, but his paper makes it clear that there were differences of opinion on the nature of what an antiskid system did. Strien himself did define what he thought constituted an antiskid device: "An antiskid device starts to work only after the beginning of the locking process, and can be termed only then as functionproof and as 'locking-prevention device,' when it diminishes the braking effect in such a rapid and decisive manner so as the wheel or wheels rotate 'dosedly-braked' until the vehicle comes to rest."[6] Strien, who had been one of the pioneers of proportional braking system design in the 1950s and who moved into the field of antiskid in the 1960s, had an interest in differentiating the projects. Within five years his definition of antiskid devices, which limited the term to those machines that attempted to prevent locking, became the standard usage. In fact, brake-proportioning devices often did reduce skidding. The reason for this is simple and within the experience of most drivers. As a vehicle comes to an abrupt stop, the proportion of weight carried by the front wheels increases. Conversely, the load on the rear wheels lightens, increasing the likelihood that the rear wheels will lock up. Whether a wheel locks is influenced by the speed of travel, the friction between the road and the tire, friction between the brake shoe and disc or drum, and the load on the wheel. Heavier loads make wheel lock-up less likely. Brake-proportioning valves reduce the pressure to the rear brakes as the load shifts forward, making it less likely that the rear wheels will lock. As explained earlier, it is more danger- ous to have the rear wheels of a car lock, since this causes the car to spin 180 degrees. Still, Strien's argument is perfectly sensible: brake- proportioning devices, while improving braking performance, do not perform the same function as antilock systems. However, even dis- qualifying brake-limiting systems as antilock devices did not alleviate all disputes over what constituted the first production antilock system.

AMERICAN CLAIMS

In the 1960s and 1970s several companies claimed to bring out the first production antilock braking device. The first production systems de- signed solely to prevent skidding, excepting the British Maxaret system made by Dunlop, were American. All three major American automobile manufacturers devoted considerable resources to antiskid R&D in the 1960s. All three developed functional antiskid prototypes, and GM was targeted by Ralph Nader in *Unsafe at Any Speed* for developing an antiskid system and withholding it from the market.[7]

It is difficult to imagine a scenario in which GM would have anything to gain by withholding a system from the market, and GM did introduce a system in the early 1970s for the Cadillac El Dorado. In the early 1960s GM engineers did produce a simulator which allowed more advanced testing of antilock systems than either Ford or Chrysler had.[8] These antiskid devices failed to meet one critical performance standard that GM had set: they did not reduce stopping distance, and for this reason GM was reluctant to release the system on the market. Neither Ford's nor Chrysler's system reduced stopping distance either, but the three companies set different performance criteria. Ford and Chrysler devel- oped antiskid options for braking systems on their high-end, luxury automobiles in collaboration with two major American brake suppliers, Kelsey-Hayes with Ford and Bendix with Chrysler. Their intention was to introduce the system initially at the high end and then, through improved production techniques, make the systems less expensive and filter the technology down to less and less expensive cars. The Ford system debuted first, in the fall of 1968, and the more complicated Chrysler system was introduced in the fall of 1970.

While all of the early antiskid devices had some conceptual similarities, the functional differences were quite significant, allowing each subsequent design to be presented as substantially new. In the eyes of the engineers, how these systems worked defined what they were. This relationship facilitated the complicated claims of priority in several cases over a decade. The antiskid devices of Dunlop and the American manufacturers were all of a similar type, and all drew on the experience of brake testing to design devices which could detect skidding as a factor of changing velocity and modulate pressure to a brake system. The sensor and signal-processing technology of Dunlop's Maxaret and the Ford and Chrysler systems used the electronics technology of the 1960s. These devices had more in common with the brake-testing machinery that preceded them than with the antilock braking systems yet to come.

Kelsey-Hayes's and Ford Motor Company's System, 1969

In a paper detailing the development of their antiskid system, R. H. Madison of Ford Motor Company and Hugh Riordan of Kelsey-Hayes began with a surprisingly lengthy discussion of the important developments in automobile braking. Clearly their intent was to position their invention, which they called the Sure-Track Brake System, among these innovations. They noted the introduction of several improvements in braking systems, leaving only the problem of skidding to be solved. According to Madison and Riordan, the innovations leading to the attention on skidding included the following:

—Braking systems which engaged all four wheels
—Hydraulic, instead of mechanical, actuation
—Power-assisted braking systems
—Improvements in brake lining materials
—Disc brakes
—Dual master cylinders
—Interior system-failure warning lights[9]

The article claimed antiskid devices and, by implication, the Sure-Track system, as the next development in this list of successes. But Madison and Riordan were careful to establish their own and their companies' credentials in the research and development of new braking systems. While innovative brake design had been perceived as a largely European field, Madison and Riordan pointed out that Ford had been involved in attempts to develop antiskid devices since the early 1950s. Ford's first project had been to attach an airplane antilock unit to a 1954 Lincoln, a project remarkably similar to Lister's, performed at roughly the same time. Ford's experience was the same as Lister's, too: the unit vibrated the automobile uncontrollably. Kelsey-Hayes began work on antilock systems for automobiles in 1957, after being one of about a dozen companies to design and produce antilock systems for aircraft.[10] The project which led to the Sure-Track also involved research by Hydro-Aire (another aircraft antilock system producer), Texas Instruments, Teledyne, the Cornell Aeronautical Laboratory (Milliken's and Whitcomb's institute), Battelle Institute, and Rensselaer Polytechnic Institute.

A significant portion of Madison's and Riordan's article dealt with the criteria these cooperating organizations used in development, which formed the primary distinguishing characteristics between different systems. These criteria often facilitated the priority claims each producer made. In all antilock systems one principal defining characteristic was the number of wheels in which skidding was monitored and prevented. Obviously, systems that engaged only two wheels were simpler and cheaper, but the producers of four-wheel systems argued that the performance offered by the lower cost two-wheel systems did not justify their cost; they were cheap, but a poor value.[11] For an enhancement of an already functional braking system, performance criteria were more important than price, they argued. However, Kelsey-Hayes's and Ford's engineers argued that the time it would take to develop a four-wheel system might relegate Ford to second place in the race to produce the first American, production-line antiskid device.[12] Since Kelsey-Hayes sold antilock braking systems for aircraft, they knew that the cost was based on the number of wheels involved ($1,500 per wheel for aircraft). Doubling the number of wheels would roughly double the cost, making the system prohibitively expensive. This was ample evidence to convince Ford's engineers to commit to a two-wheel system. In addition, using an antiskid device only on the rear wheels eliminated the driver's ability, as well as his need, to countersteer during panicked braking; the vehicle's front wheels would lose steerability if they locked up, but this would result in the car's continuing in the same line in which it had started.[13] Agencies studying auto safety found that during an emergency stop drivers usually lacked the skill necessary to countersteer the vehicle. Preventing the rear end of the car from skidding would prevent rotation, so the driver would have no need to countersteer.

Once Kelsey-Hayes and Ford had committed to a two-wheel system the only other major decision critical to the development process was which of the existent power systems in the automobile to use. Kelsey-Hayes and Ford were working on a system that was to be added onto an existing braking system, and this limited their choices of power supply. They chose to use the automobile's existing vacuum power supply, a decision shared by many of the early antilock developers.[14] Madison and Riordan did consider other possibilities, including using electrical power from the car's alternator, but dismissed this option as too slow. They also discussed using the hydraulics of the power steering system or automatic transmission, but decided that this needlessly complicated the system by requiring it to interface with multiple component groups. European manufacturers had to make similar choices about power, but they dismissed power steering and automatic transmissions because these were quite uncommon on European cars.

After engineers had committed to a two-wheel, vacuum-powered system, the other criteria were set according to the performance standards of Ford and Kelsey-Hayes. Madison and Riordan detailed these criteria in their article. The Sure-Track system was designed to operate at all speeds and on all roads with friction coefficients from 0.05 (ice) to 1 (a road surface so sticky it is purely theoretical). Ford and Kelsey-Hayes did not specify that their system had to shorten braking distances, but lengthening the stopping distance from that which could be achieved by manually pumping the brakes was undesirable. Ford required that the Sure-Track system contain a failure mode that would leave the braking system in normal operation if the antiskid device failed for any reason. Sure-Track's control system kept the sensoring equipment, measuring only angular velocity. An electronic control unit calculated and compared deceleration and acceleration. The system could not require any new, additional, or peculiar maintenance over that necessary for normal brakes. Engineers designed Sure-Track so that existing equipment would not be affected by it. More important, they assumed that the system would not require or expect drivers' behavior to change at all.[15] In other words, the Sure-Track was an add-on component, designed to work with a number of different systems in use in Ford and Lincoln vehicles. Furthermore, Sure-Track was truly black-boxed; it had to operate without the driver's knowledge. Sure-Track offered the driver greater stability rather than steerability or an improvement in stopping distance. None of the systems of the 1960s shortened stopping distances, so Kelsey-Hayes and Ford sacrificed this criterion in order to bring the system to the market as early as possible.

Sure-Track worked only on the rear wheels of a vehicle, reading the angular velocity of the rear axle using floating sensors and sending a pulse signal to a processor under the glove compartment. The processor converted and compared the angular velocity of the axle with the speedometer reading of linear velocity. Acceleration and deceleration were not measured directly, but were calculated from velocity readings over time. If the rear axle decelerated more quickly than the vehicle, then the vacuum-powered rear brake released and reapplied itself four times per second.[16] This would result in a braking action even rougher than that of the Maxaret, which could pulse six times per second. In this and many other ways, the design of the Sure-Track was more similar to the Maxaret than to the antilock systems developed subsequently. Both the Maxaret and the Sure-Track were add-on components, not new braking systems. While the Maxaret system could be used on all four wheels, it also did not improve braking distances on most surfaces. In both cases, the principal improvement was in the retention of stability, preventing the car from spinning 180 degrees when the rear wheels locked.

Despite Ford's aim of mass production, the Sure-Track was never offered on any models other than the relatively expensive Lincoln Mark III and Ford Thunderbird. It was never widely used, and its add-on nature meant it was a device and not a system. The Sure-Track clearly fell in the same category as the Maxaret; it was important to the development of the *promise* of antilock braking, but not a market success in its own right.

Chrysler, Bendix, and the 1971 Imperial

Chrysler's system was more ambitious than Ford's. Chrysler began work on skid control in 1957 and was focused on production designs by 1966. The Sure-Brake was introduced in the fall of 1970 on the 1971 Chrysler Imperial.[17] Chrysler made different design decisions than Ford did, and the extra two years of development time meant a more sophisticated four-wheel design with better electronics. Bendix's and Chrysler's engineers fell into the category of those who believed that performance of a two-wheel system was simply unacceptable. While a four-wheel system cost more, it also offered the driver more control.[18] When all four wheels were prevented from locking, the driver retained the ability to steer the car as well as maintain stability. Chrysler also wanted to shorten the distance it took for the car to come to a stop over the distance a skilled test driver could achieve by pumping the brakes.

Aside from deciding to design a four-wheel system due to its performance advantages, engineers at Chrysler and Bendix had several other design decisions to make in order to initiate a testing scheme for the Sure-Brake.[19] They had to commit to the type of sensors, mechanical or electrical, and they chose electrical. With electrical sensors the device would require some kind of electronic control, which at this stage was analog. Acceleration or velocity could be measured at either the wheel or the drive shaft. If they measured speed or acceleration at the drive shaft, would it be through the speedometer cable or through the differential housing? These questions were discussed but proved moot once they decided to pursue designing sensors at each wheel. Chrysler's Sure-Brake would measure angular velocity at each wheel, unlike the Kelsey-Hayes and Ford device that measured it at the axle. Like the engineers at Ford and Kelsey-Hayes, they decided for simplicity's sake to use the vacuum booster to provide a power source to the system. Last, they had to determine the degree of integration into the braking system. Engineers at Chrysler and Bendix wanted to know whether greater integration would lower or raise the cost of the system. In the end, the increased complexity of integrating the antiskid system into the braking system itself overrode any cost savings due to eliminating duplicate parts. Decisions regarding these five initial conditions were made, based on test results at the Chrysler Proving Grounds.[20]

This system was presented to the SAE in a paper given at the Automotive Congress in 1971 by J. W. Douglas of Chrysler and T. C. Schafer of Bendix. The ground rules for the Sure-Brake system were as follows:

—Good steering control and stability had to exist under all braking conditions.
—Stopping distances had to improve on those a skilled driver could achieve by pumping the brakes.

—The braking system was to be unaffected by the weight distribution of the vehicle.

—The Sure-Brake had to be installable as an option on the 1971 Imperial.[21]

Chrysler and Bendix were independently developing systems that matched these criteria, and in 1969 Chrysler entered into an agreement with Bendix to bring the Sure-Brake system into production in time for the 1971 model year. Like Kelsey-Hayes and Dunlop, Bendix had been one of the developers and producers of antilock systems for aircraft in the early 1950s. Possessing similar experiences in inventing antiskid devices for aircraft may account for some similarities between the Sure-Brake, the Sure-Track, and the Maxaret systems.

Still, there were several novelties to the Chrysler system. The sensors were different from those used by either Ford or the Maxaret, although the Maxaret too used wheel sensors. The front sensors of the Sure-Brake system were magnetic, while on the rear the sensors were spring-loaded gears. Engineers designed the rear sensors to accommodate large displacements between the rear axle and the brake. The sensors measured angular velocity and sent this information in pulses to a logic controller. The logic controller differentiated wheel speed by time to calculate deceleration. The system had a preset threshold but was also adaptive within a preset range. When deceleration reached 1.6g, the logic controller would store the angular velocity of each wheel. If wheel speed dropped 5 percent when deceleration was below 16 feet per second per second or 15 percent if deceleration was greater than 16 feet per second per second, brake pressure to the slow wheel would be reduced. Once the wheel started to accelerate at a rate greater than 0.2g, the pressure reduction to that wheel would stop and pressure would build up again. The Sure-Brake system could cycle at six times per second, about equal to the Maxaret. This would certainly be perceived by driver and passengers as a jerky and rough motion.

Using a device like the RRL's fifth wheel apparatus, the Sure-Brake produced the improvements in stopping distance shown in table 1. In other words, the Sure-Brake worked: it shortened stopping distances on the surfaces where control had been more difficult before. In addition, both steerability and stability were maintained by the driver. The Sure-Brake performance standards were higher than those demanded by Ford and Kelsey-Hayes and outperformed the Sure-Track. Whether the

Table 1 *Comparison of Stopping Distances (in feet)*

	Locked brakes	Brakes pumped by skilled test driver	Sure-Brake
60 mph dry concrete	159.2	192.5	165.3
60 mph wet concrete	199.8	215.2	176.5
ice	424.5	521	417.3

performance standards were such that consumers would respond was yet to be seen.

Once Chrysler and Bendix engineers knew that the system's performance justified continued work on it, they had to face problems of reliability. The system worked well enough under laboratory and relatively fastidious test track conditions, but once on the road the Sure-Brake would hardly have such a pampered existence. The first real-world problem engineers encountered was the system's intolerance for salt, commonly used to melt ice on road surfaces. The presence of salt caused current to be induced from the car's electrical system into the highly sensitive "milli-volt circuits of the wheel control system."[22] Unfortunately, salty roads were a condition the system was certain to face, so the antiskid system needed insulation to prevent the electrical system of the auto from bleeding into the brake control system. In addition, there were concerns about the effect of electromagnetic interference from radio and TV towers and from the radio in the car. For two months engineers tried to figure out why the system occasionally failed without warning at night. They finally discerned that failures occurred because the nighttime test drivers played the radio while driving. In addition, the Voice of America radio tower in Cincinnati caused random failures. As in the case of electrical interference, the solutions were found in better insulating the system.

Chrysler debuted the Sure-Brake system on the 1971 model Imperial, but found that American customers were, at best, lukewarm to the optional safety device. The system became an exemplar of the idea that safety still did not sell, even after the publication of Nader's *Unsafe at Any Speed*. The relationship of American consumers to automotive safety options is a long and complex story, and until the 1990s one without a clear message. Despite the fact that Chrysler and Ford sold very few of their antilock systems, cars were becoming steadily safer. In *The American Automobile Industry*, John Rae offers statistics showing the decline in deaths per 100 million motor miles from 8.82 in 1947 to 5.36 in 1965.[23] The passage in 1966 of the National Traffic and Motor Vehicle Safety Act ironically drew some attention away from consumers' choices in safety equipment by inadvertently creating the widespread assumption that if a safety option was necessary the government would require it. Some Americans saw the creation of the National Highway Traffic Safety Administration as an intrusion, forcing up the price of cars with unnecessary additions, such as safety belts, which NHTSA mandated as standard equipment in 1968. That sector of American drivers was best represented in 1974 by the outcry over and repeal of the NHTSA's requirement that seat belts be fastened before an automobile could be started (safety interlocks).[24]

The attitude of American automobile manufacturers was clearly at odds with public health investigations that started the search for better braking systems in the 1950s. Jameson Wetmore claims that it became clear during the Ribicoff hearings investigating the Nader report in 1965 that "carmakers did not envision the driver as a member of the public. To them, he (and drivers were largely seen as male) was a consumer, a person to whom they wanted to sell cars. Thus, the prospect of making numerous and expensive changes to automobile interiors caused manufacturers to be afraid of alienating their clientele. . . . The manufacturers contended that it was the consumer's responsibility to buy such items just as they bought air conditioners, vinyl roofs and high-horsepower engines."[25] If, as in the case of the Sure-Brake system on the Imperial, consumers did not want the systems, then Chrysler had a responsibility to offer them the options they did want. Spending an extra $800 to get a system which merely improved braking distance and handling on wet roads proved a hard sell to the American public in the 1970s.

There is another lesson to be drawn from American manufacturers' attempts to produce a marketable ABS. This episode shows the relative isolation of the American engineers from the knowledge community. European engineers were the most active participants in the ABS knowledge community. They were the ones pushing the community to publish in different languages and to distribute their research findings widely. Americans, on the other hand, rarely published outside the

Society for Automotive Engineers and their own professional cohort. Germans also published in the SAE literature, so while German engineers regularly read most of what American engineers wrote, American engineers, especially those working in industry, read only a minority of what German engineers wrote, since there were more articles on ABS published in the *Automobiltechnische Zeitung* than anywhere else. One reason for this was that American companies required tighter non-disclosure agreements from their employees. The fact that their employees could not participate in the community as completely as employees of Teldix, Daimler, or Teves lessened the flow of knowledge. American engineers simply knew less about the work going on in Europe, and they overlooked one of the major resources that their knowledge communities provided. The ultimate failure of American systems from Chrysler, Ford, and General Motors cannot be attributed to the lack of skill of their participants so much as to the unwillingness of the corporations to see the value of knowledge exchange.

seven

Eines ist sicher!

Successful Antilock Systems in West Germany

The Maxaret, Sure-Track, and Sure-Brake were components of
the 1960s. They all repeatedly measured instantaneous angular
velocity at a vehicle's wheel or wheels, instead of measuring
either angular or linear acceleration or deceleration. They were
all designed to operate in an already existent braking system. In
this sense, none of these three devices constituted a system in its
own right; each was an add-on component of a preexistent and
previously designed braking system. Moreover, none of these
devices really succeeded in establishing the market for antilock
braking. While there could have been further development of
any of these components, the emergence of vehicle emissions
standards and control in the 1960s limited the degree to which
the Maxaret, Sure-Track, or Sure-Brake could have been improved. Emissions standards developed independently from
antilock technology, although they had an unintended, but critical, effect on antilock specifications. Exhaust and emission control reduced a car's vacuum power, and all of the antilock systems of the 1960s relied on vacuum power to assist the braking
system's hydraulics. Catalytic converters, the chief emissions
reduction technology of the 1960s, reduced vacuum power by
nearly 40 percent. Although this choice was never put to consumers in this way, given the mechanics of the antilock systems
which were available in the 1970s the American car-buying
public could have either antiskid devices or catalytic converters.
But consumers did not make this choice; first states did, then
the U.S. federal government, then European governments fol-

lowed suit. In a good example in the history of technology showing why the technical details of the way devices work matter, engineers suddenly faced a real limit on how to improve vacuum-driven antilock systems.

However, not all antilock devices were vacuum-driven. American manufacturers had chosen to develop vacuum systems in part because they found electronic systems unreliable. Not all engineers shared this view; members of the knowledge community who also worked in avionics were the most likely to trust electronic systems, which were much more sophisticated and reliable than automotive electronics. By choosing to require catalytic converters, the U.S. government created an advantage for antilock systems that did not rely on vacuum power; ironically, these were the systems being developed by the West German companies Teldix, Bosch, and Teves.

In addition, the oil crisis of 1973–74 reoriented R&D in the automobile industries of the world, making improvements in fuel efficiency a greater priority than developing antiskid devices. This reorientation was particularly advantageous to Bosch, fresh from the success of mass-market electronic fuel injection with the Druck-Jetronic (available on the 1967 Volkswagen 1600TL), which replaced the traditional carburetor and both increased fuel economy and decreased emissions.[1] Furthermore, in the early 1970s fuel injection systems were developed at Bosch, which also worked with catalytic converters (KE-Jetronic). Bosch's previous experience with fuel injectors was significant because it meant Bosch had already mass-produced relatively cheap automotive electronics in a reluctant industry. Now that manufacturers expected electronic fuel injection to be cheap and reliable, Bosch could simply create a similar strategy for getting them to accept and install ABS.[2] With the phrase "Eines ist sicher!" Bosch used this approach to promote ABS. The phrase has a double meaning in German; idiomatically it means "One thing is certain," but *sicher* also means "safe."[3] By the mid-1970s Bosch was certain that their system offered precisely the performance criteria that would overshadow all the market failures in antilock systems. But before the system hit the market in 1978, there was a lot of R&D work yet to do and an important corporate collaboration between Teldix and Bosch to establish.

The passage of emissions regulations and the oil crisis delayed the production of new antiskid systems in the United States. Surprisingly, the crises of the mid-1970s in the automobile industry did not significantly slow or scale down antiskid research in West Germany, only the pace of introducing new systems on the market. The slower pace of research in the United States reflects the failure of the Kelsey-Hayes and Bendix systems as well as the stagnating U.S. economy. But around the world, after a series of unsuccessful antilock devices hit very limited markets in the 1960s, almost a decade passed before the introduction of the next generation of antiskid devices. The stalling of the market for new ABS could have been disastrous for the engineering community, but in fact it meant that the second round of antiskid devices were far more sophisticated than the first. One might even argue that the problems of the early and mid-1970s prevented a second, not-quite-ready generation of machines from ever appearing on the market. Bosch's first system was even called ABS-2. There was never any mass-produced ABS-1 system; this designation refers to the system that the Teldix engineers Heinz Leiber, Hans Jürgen Gerstenmeier, Wolf-Dieter Jonner, and others developed in the 1960s. In the history of ABS, it seems that the devices of the late 1970s were really a third generation, after the first-generation Maxaret, Sure-Track, and Sure-Brake systems and a second generation that never appeared commercially.

Because their immediate predecessors had nearly faded from memory, the producers of these later devices were able to claim that their devices were the first antilock braking systems. Antilock system producers of the 1970s wanted to distance themselves from the less than successful systems that preceded them. Promoting the reliability of the new systems in the 1970s required explaining to car manufacturers that they were meeting different standards than their predecessors had; they wanted their customers to see these systems as substantially different from those offered earlier by Dunlop, Bendix, and Kelsey-Hayes.[4]

More important, in the 1970s engineers made great improvements in electronics technology, making signal processing more reliable, faster, and less expensive by the end of the decade.[5] The systems of the 1960s all raised questions about the reliability of their electrical systems.[6] The antilock braking systems of the 1970s relied more heavily on solid-state electronics than their predecessors had, but by 1978 electronics proved to be their strength instead of their weakness. While automobile manufacturers and their suppliers waited out the economic downturn of the early 1970s, electronics technology caught up. In their study of the semiconductor industry, *A Revolution in Miniature*, Ernest Braun and

Stuart Macdonald discuss the sorry state of electronic devices in automobiles in the late 1960s and early 1970s: "In the United States, safety and health legislation in the seventies virtually forced the adoption of more electronics. . . . Such electronics appears to have been unpopular with manufacturers and users alike for the economic depression of the mid seventies put much of the legislation in abeyance and the electronics frequently disappeared. . . . The automobile manufacturers seem to have assumed that semiconductor electronics in cars was not the sort of thing that would make their product more attractive and, hence, more competitive."[7] Manufacturers' trust in sophisticated automotive electronic braking systems also declined because of misfortunes such as the passage in 1975 of Federal Motor Vehicle Safety Standard 121, which required antilock systems on the air brakes of semitrailers, to be phased in to prevent jackknifing accidents that often occurred as a result of skidding. In an unlikely marriage of interests, the Teamsters Union and truck manufacturers banded together to oppose FMVSS 121; it was rescinded in 1978.[8] Manufacturers and drivers alike were highly skeptical of these new systems. The negative publicity of the problems related to truck antilock systems frightened off the automobile manufacturers until they were convinced that antiskid devices showed marked improvement over skilled drivers. This informal criterion was difficult to meet, because it required a new approach and new components. As a result the knowledge community had to reorient itself to these new demands for performance and reliability, and this often meant changes in the composition of the community as well as its relationship to the previously remote automobile producers.

GERMAN PARTNERSHIPS: TELDIX, ROBERT BOSCH, AND DAIMLER-BENZ

In the 1960s and early 1970s a small company in Heidelberg, West Germany, strategized about how to overcome automobile manufacturers' anxieties about automotive electronics. Teldix, GmbH, was formed by Telefunken and Bendix to produce avionics for the American F104 Starfighter supersonic interceptor military plane, produced between 1958 and 1967.[9] In 1973 Robert Bosch, GmbH, purchased a 50 percent share of Teldix (what amounted to the Bendix interest), largely to infuse Teldix with the knowledge and financial resources necessary to develop production-scale electronics.[10] Both Teldix and Bosch had previously been developing antilock systems, Teldix since 1963 and Bosch since 1969. Thus both companies had lengthy R&D experience with automotive electronics. In developing gyroscopic avionics systems for the Starfighter, Teldix engineers were experienced in working on the interface of electronic control and mechanical components. The Robert Bosch corporation was a venerable company founded in 1886 by Robert Bosch, who built the company's fame and fortune on his invention of the spark plug in 1902. Today, Bosch is an incredibly diverse, multinational, billion-dollar corporation producing everything from video surveillance systems to dishwashers, but its origins were in the automotive industry. Automotive technologies remain the heart of its research and development operations at its headquarters in Stuttgart, Germany. In the 1960s, Bosch was riding high on its electronic fuel injectors, which posed problems of a similar nature to ABS, requiring adaptive electronic control of a hydraulic system. As mentioned earlier, Bosch was already mass-producing electronic fuel injectors when ABS attracted the company's attention in the mid-1960s. Both Bosch and Teldix took a systems approach to designing electronic components and control into already complex mechanical systems. Both companies were familiar with the relationship between safety, system failure, and reliability in their customers' minds.

In many ways, Bosch and Teldix brought a different kind of expertise to antiskid research than Bendix and Dunlop had. Those companies worked from their strength in designing and developing braking systems. Bosch and Teldix were not involved in the design of braking systems; their work focused on electronic antilock devices that could be added on to any number of different manufacturers' systems. What they delivered was expertise in electronic control. Their electronic orientation would serve them well in combating the anti-electronics prejudices of automobile manufacturers and the public in the 1970s. With their involvement, and the attention of large automobile firms like Daimler and BMW, the knowledge community once again recast its central questions, now adding questions about the electronic control of braking into the mixture of interests, challenges, and rewards.

In 1973 the beta version of Teldix's ABS-1 contained over a thousand semiconductors in a 40x20cm case and was too expensive and fragile to be commercially produced.[11] By the time ABS first appeared on the market in 1978 this very large number of transistors had been replaced with purpose-designed, mass-produced, digital large-scale integrated circuits.[12] The five-year gap in commercial introductions of new braking systems, attributable to the oil crisis and general economic malaise of the early to mid-1970s, changed the nature of automotive electronics. It has already been established that the SAE credited engineers from Teldix and Bosch for the development and market introduction of ABS with the Sperry Award. But the SAE and other associated engineering societies did not present the award until 1993, by which time it was clear that the electromechanical system Teldix and Bosch had brought to market in 1978 had stood the test of time. Consequently, Teldix's claims of innovation and priority from the 1970s were no different from the claims made by Chrysler or Dunlop at the time. The system that Teldix started developing in the 1960s was not self-evidently better; instead, the system's perceived superiority was consciously constructed over nearly a decade using the communication tools at hand: a continuous flow of presentations at auto shows and articles in noteworthy engineering publications. The audience for Teldix's claims of superiority and priority was other engineers, sometimes those working for competing companies, sometimes those working for the automobile manufacturers. Teldix depended on a technically knowledgeable audience who could recognize that its design confronted the difficult design issues that Ford, Chrysler, and Dunlop had failed to solve. Teldix's system would embrace all three design goals: steerability, stability, and the one goal no existing antiskid device had achieved, a shorter braking distance.

Before ever beginning work on antilock braking systems, Teldix was known for the invention of an electronically controlled hydraulic valve with very low inertia that offered an extremely fast response time under very high pressure.[13] Such a valve was a completely new component in the automobile industry, where hydraulic systems, though common to every vehicle, were far less sophisticated than in aviation. Manufacturing the Teldix valve took a higher level of precision machining than was commonplace in the automobile industry. The valve was small and its movement very sensitive, so that very little force was needed to open and close it; this is what *low inertia* means. Because it was low-inertia, the valve could pulse sixty times per second (compared to four or six times for the valves used by the Maxaret, Sure-Brake, and Sure-Track). With this speed, engineers gained much more flexibility in controlling the amount of retardation a brake system generated. In hindsight, it's clear that high-speed valves were the technological key to improving braking efficiency. Using technology that made greater efficiency realizable, engineers at Teldix rethought the relationship between efficiency and wheel slip, exactly the problem that concerned engineers constructing vehicle dynamics models. Teldix's valve facilitated a marriage between antilock systems and vehicle dynamics models; previously distant parts of the knowledge community came together.

There are two different accounts of how Teldix and Daimler struck a partnership in 1969 for the development of ABS. According to Jean-Pierre Gosselin's article on the history of ABS, Teldix's valve caught the attention of engineers at Daimler-Benz and drove the partnership between Daimler and Teldix. Accounts from Teldix engineers tell a slightly different story; they claim that they initiated the partnership with Daimler.[14] Teldix needed test vehicles and facilities for the first antilock system prototypes and contacted Daimler, located 30 kilometers away in Stuttgart, making it the closest large auto manufacturer to Heidelberg. Since the relationship between Daimler and Teldix was arranged after Teldix had produced a prototype system, the Teldix account seems more likely. In either case, a relationship was created in the late 1960s between Teldix and Daimler to collaborate on an electronic antilock braking system.

THE TELDIX ABS

The antilock R&D project began at Teldix in 1964 under the direction of Heinz Leiber, who was Teldix's director of development. Leiber had been trained as a precision mechanical engineer and had worked on aviation navigational systems before antilock research began.[15] Under his leadership, Teldix assembled an R&D team that included Wolfgang Limpert, a mechanical engineer. Wolf-Dieter Jonner headed up the testing arm of the Teldix team. In 1967 Leiber added an electrical engineer, Hans Jürgen Gerstenmeier, to work on the electronic control systems. Between 1969 and 1975 several other engineers were brought on board, including Armin Czinczel. All these engineers transferred with the project to Bosch in 1975.

The Teldix team concentrated on producing a device that had four design objectives. Teldix required the system to (1) prevent *any* of the vehicle's wheels from being 100 percent locked up. This would, in turn, (2) shorten stopping distance on both dry and wet pavement,

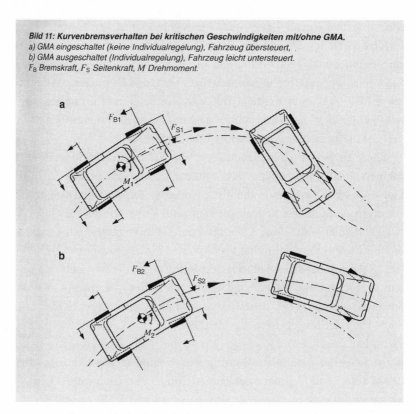

Bild 11: Kurvenbremsverhalten bei kritischen Geschwindigkeiten mit/ohne GMA.
a) GMA eingeschaltet (keine Individualregelung), Fahrzeug übersteuert,
b) GMA ausgeschaltet (Individualregelung), Fahrzeug leicht untersteuert.
F_B Bremskraft, F_S Seitenkraft, M Drehmoment.

2 Diagram showing sideslip during braking on a curve, with and without yaw moment delay (GMA). (a) GMA switched on (no individual control), vehicle oversteered. (b) GMA switched off (individual control), vehicle lightly understeered. F_b = braking force, F_s = side force, M = torque.

English translation of German text; reprinted from Robert Bosch GmbH, *Bremsanlagen für Kraftfahrzeuge* (Stuttgart: Robert Bosch GmbH, 1994), 47.

(3) maintain lateral force while braking in a curve (figure 2), and (4) allow the driver to retain directional control of the vehicle.[16] Teldix's goals were directly influenced by the work of vehicle dynamicists, such as Mitschke in Germany and Milliken and Whitcomb in the United States. Leiber and his colleagues often claimed that their system was derived from theories of tire friction, a unique claim among antiskid developers.[17] Their design goals were aimed at realizing what was theoretically possible according to theories of vehicle and tire dynamics, rather than seeking simple improvements over existent braking system technology.

By 1968 Leiber was writing articles that emphasized the differences between Teldix's research and other antiskid projects. He specified that the unique aspects of Teldix's antilock system were its fast response time, due in large part to the improvements in valve technology unique to Teldix, and its ability to adapt itself to changing road surface conditions.[18] Teldix presented the designs, referred to as ABS, as the first adaptive antilock system.[19] Adaptive control systems were all the rage in electronics in the 1960s; for braking this meant that the system was not triggered by reaching a certain *g* force, but had a much more sophisticated method of measuring and calculating whether the wheels were slipping and when to pulse the brakes. For example, Chrysler's Sure-Brake *calculated* deceleration (*g* force) from angular velocity over time.[20] It pulsed the brakes when a threshold of 1.6*g* was reached (that is, 60 percent more force than gravity, which is 1*g*). Teldix's ABS *measured g* directly using a gyroscopic aviation system to measure deceleration; then it calculated wheel slip from deceleration. The brakes were modulated when it calculated a change in the friction between the road and any wheel.[21] In this way, ABS adapted itself to the road conditions; it did not engage at any set point and it did not engage at any set frequency or brake pressure. The system responded differently to different road conditions by directly measuring more variables. The mathematical model Teldix was using was far more sophisticated than the systems of the 1960s, which were looking only at comparative velocity. This sophistication came from Teldix engineers' knowledge of vehicle dynamics models and the mathematically complex ways they related angular velocity, angular and linear deceleration, wheel slip, friction, and braking efficiency. The knowledge community was at work here, and the appearance of Mitschke's mathematical models in the Leiber design shows interdisciplinary know-how moving through the knowledge community. According to Mitschke's models, braking efficiency had to be in the 80 to 90 percent range if stopping distances with ABS were to improve on those offered by partially locked wheels.[22] For Leiber, stopping distance had to be better with ABS than without, period.

More than any other contemporary system, Teldix's antilock system was designed to improve control of the vehicle during both braking and cornering. While other antiskid devices aimed at preventing skidding, Teldix's approach was to *regulate* skidding. In order to improve stopping distances, the antilock system had to tolerate some skidding because braking was most efficient with some minor wheel slip. In addition to maintaining steerability and stability, Teldix engineers wanted to produce a system that prevented the car from sliding off the road while turning, the so-called sideslip problem. This meant the system had to accommodate differences in speed, acceleration and deceleration, and the coefficient of friction between the right and left sides of the vehicle. Leiber and his collaborator Limpert explicitly asked how much slip was enough to increase braking efficiency and how much was too much, causing the driver to lose control.[23] The amount of slip, expressed as a percentage where 0 percent meant no slippage of the wheel against the pavement and 100 percent meant the wheel had completely locked up, required an understanding of rubber tire behavior, especially in the areas of friction and the dynamics of lateral forces and sideslip. If the Teldix system could optimize braking efficiency then it would offer a major improvement in braking technology, which could justify its higher cost. In 1971 Leiber claimed that the desirable range of slip was 15 to 25 percent, depending on road conditions.[24] While other antiskid producers worried about cost and making their machines available to large numbers of drivers, Teldix's strategy was built on pure performance, a perspective Teldix brought from the avionics field. Making the Teldix antilock system mass-producible and therefore affordable was a second developmental stage, not an initial design consideration.

Leiber's emphasis on performance eliminated many of the design considerations crucial to other projects. He admitted that his system would not be easily adapted to any vehicle; it would have to be model-specific since the dynamic behavior of different models of cars was not the same, and algorithms used by the electronic control unit would have to be unique to different models.[25] Leiber and the electrical engineer Gerstenmeier first focused on making an analog system, with digital control as a goal for the second generation. From the beginning, Teldix planned that its antilock system would use digital logic circuits, but in the late 1960s integrated circuits were neither reliable nor plentiful enough for use in automotive components.[26]

Engineers at Teldix did not rely solely on wheel speed as a measure of slip. All other antiskid systems measured angular velocity and differentiated it by time to calculate deceleration. Teldix's system still measured wheel speed, but also used decelerometers developed for brake system testing. This gave the engineers options in how to measure imminent skidding and allow some wheels to slip before modulating the brake pressure. Their ABS could compare wheel speeds with vehicle speed, with each other, and decelerations across these parameters.[27] In addition, ABS could tell whether a wheel had stopped spinning completely and react accordingly.

Engineers tried several different methods of measuring speed and slip to create a system that functionally calculated the percentage a wheel was slipping. This was a great benefit in braking on surfaces where each wheel experienced a different coefficient. Teldix and Daimler built a track with unequal coefficients on friction on each side to assess the stability of the vehicle.[28] For engineers at Teldix, this was not a luxury; all cars operated under conditions in which the coefficient of friction varied between right and left wheels because "roads vary, tires vary, and loads vary."[29] The Teldix system could accommodate a difference of 0.8 in the coefficient of friction between the right and left wheel with no loss of stability.[30] Teldix's more sophisticated means of measuring deceleration led to a more precise measurement of slip, allowing the system to maximize braking efficiency and therefore reduce the stopping distance.[31] This precision meant that every model of car needed a slightly different algorithm, depending on weight distribution, wheelbase, braking system, wheel size, and the interrelationship of these factors. Teldix's system of measuring deceleration was truly a system: the parts did not function individually as components, but uniquely as a whole package.

By 1971 Teldix had worked out a collaborative agreement with Daimler-Benz, and the antilock braking system, termed ABS, was being tested on Daimler's test tracks. In this first generation Teldix used both spring-damped acceleration sensors and pulse generators on the front wheels and only pulse generators on the rear wheels (figures 3 and 4). These components sent signals to a circuit board in the trunk of the car, which engaged a system of high-speed electromagnetic valves.[32]

The valves regulated the system; they did not simply modulate the brakes or toggle them on and off. Valves could maintain five different options: a fast or slow pulsed pressure rise, constant pressure, or a fast or slow pulsed pressure drop.[33] The system chose which corrective to engage based on the linear velocity of the vehicle. If the friction coefficient at any wheel jumped suddenly, the system would override the velocity range to try to prevent the wheels from locking and any accompanying instability. The coefficient of friction between tire and road could jump from 0.9 to 0.1, with no more than 200 milliseconds of

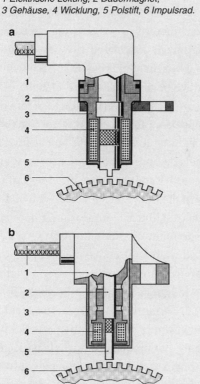

Bild 3: Drehzahlsensoren (Schnitt).
a) Drehzahlsensor DF2 mit Meißelpolstift,
b) Drehzahlsensor DF3 mit Rundpolstift.
1 Elektrische Leitung, 2 Dauermagnet,
3 Gehäuse, 4 Wicklung, 5 Polstift, 6 Impulsrad.

a

1
2
3
4
5
6

b

1
2
3
4
5
6

3 RPM sensors (cut-through). (a) RPM sensor DF2 with chisel point terminal, (b) RPM sensors DF3 with round point terminal. (1) electric wire, (2) permanent magnet, (3) housing, (4) coil, (5) terminal pin, (6) pulse wheel.

English translation of German text; reprinted from Robert Bosch GmbH, *Bremsanlagen für Kraftfahrzeuge* (Stuttgart: Robert Bosch GmbH, 1994), 57.

4 Photo of revolution sensor in place on automobile wheel. Reprinted from Manfred Burckhardt, *Fahrwerktechnik: Radschlupf-Regelsysteme* (Würzburg: Vogel Buchverlag, 1993), 279.

complete locking at low speeds. A good dry road and a new all-weather tire has a coefficient of friction of about 0.7, while black ice has about 0.1.[34] At higher speeds, 100 percent locking was avoided altogether, even under these difficult conditions. Since the Teldix system was hydraulic and self-contained, pressure siphoned off the brakes in the case of pressure drops was recirculated back into the circuit reservoir. Consequently, pressure could be modulated much more quickly and more exactly in the Teldix system than in any of its competitors. This arrangement also helped to reduce brake fade.

The Teldix ABS was a self-contained addition to a hydraulic braking system, so it did not interact with the vacuum power system that caused so many headaches to competitors. It is worth speculating that Teldix's proximity to and relationship with Bosch helped steer Teldix away from using vacuum power as early as 1967. Bosch was the leading research and development firm in electronic fuel injection and Bosch's Druck-Jetronic fuel injector was used by Volkswagen in the 1600TL starting in 1967. Clearly Bosch's interest in fuel injection stemmed from its involvement in emissions regulation. If Bosch bet its automotive future on fuel injection in the early 1960s, it is possible that Teldix also knew that vacuum power would decline with emissions-limiting devices, and therefore Leiber's hydraulic design avoided dependence on any component which would be compromised by the passage of increasingly stringent emissions regulations. In both cases, Bosch and Teldix benefited significantly in their commitment to components compatible with emissions reduction. It is also worth mentioning that neither Teldix nor Bosch was financially dependent on the success of ABS (or in Bosch's case, even fuel injection). Teldix's main business was avionics and was tied to military aircraft as well as the promising sector of satellite technologies in the 1960s.[35] Bosch was and is a multinational, multisectoral corporation, with plenty of products and markets to tend to.

The design of Teldix's system relied heavily on the work of Ramachandra Guntur and the Delft adaptive control group as well as on Manfred Mitschke's work at Braunschweig. Mitschke had built mathematical models relating deceleration and velocity to each other and to time as they influenced the behavior of a vehicle that is being braked.[36] His model, presented at the International Automotive Congress in 1970, claimed that designers of antiskid systems needed to take into account the changing behavior of the vehicle at different speeds.[37] This was true also of the process of braking a vehicle in a curve, where different wheels were proceeding at different speeds as well as different rates of acceleration or deceleration. The sophistication of Mitschke's model led to the complex array of pressure modulations that ABS offered. Mitschke's interest in antilock systems continued as the Teldix design developed. His relationship with Teldix, and then Bosch, is an example of the close relationship these two companies enjoyed with academic engineers, a relationship that facilitated the import of complex mathematical models into the design of both ABS and its control algorithms. In 1975 Mitschke produced a computer program for simulating the response of an antilock system under the conditions of emergency braking.[38] In his program and the article describing it, Mitschke demonstrated the superior performance that a four-wheel system offered in combating lateral motion, or yaw. Engineers at Bosch combined digital simulations with track testing in Sweden for the final ABS design that was mass-produced and introduced commercially.

The system achieved most of its design objectives, reducing stopping distance and allowing the driver to retain steering and stability. Their test results were as follows:

On dry concrete at 100 km/h
  With ABS: 41.8 meters
  Without: 50 meters
On wet concrete at 100 km/h
  With ABS: 62.75 meters
  Without: 100 meters[39]

In this test ABS reduced stopping distance compared to an identical vehicle without ABS. This was the performance goal Teldix strove for and one which was easily explained to both manufacturers and end users, unlike Chrysler's Sure-Brake, which failed to reduce stopping distance on the most common surface most drivers used: dry pavement.

Like their colleagues in Britain and the United States, Teldix engineers found the reliability of even the simplest electronic circuits lacking. During testing in 1972 Teldix found that the control units failed nearly 30 percent during extreme cold weather at the testing site in Arjeplog, Sweden.[40] The delay between the construction of prototypes in the early 1970s and the introduction of ABS-2 as an option on a Mercedes sedan in 1978 was caused largely by the unreliability of production electronics.

The issue of reliability threatened to end altogether the effort to produce electronic antilock braking systems. Manfred Burckhardt, Egon-
Christian Glasner von Ostenwall, and Hellmut Krohn, all brake engineers at Daimler, pointed to the multiple layers of reliability needed in a
critical system such as ABS.[41] While Daimler agreed that ABS needed to
have reliable components, it also thought that making the system reliable as a whole was a separate and distinct problem. The overall reliability of the system had to be proven to Daimler engineers. In addition, the prototype had to translate into a reliable production product.[42]
It was expensive and complicated, but also unavoidable that components coming off the production line be rigorously tested for reliability.
Also, the ABS design had to be assessed over time; just because a design
was immediately functional did not mean it would behave perfectly over
time. Was a two-year-old system as reliable as a new one? These requirements were not only difficult to meet, but also provided several places
for the system to fail. And failures were common in the period from 1973
to 1977. It seemed to management at Teldix that the problems of reliability in the mid-1970s required a company with experience in bringing
electronic control to the market. As a small-scale producer of avionics,
Teldix had little experience in producing the hundreds of thousands, or
even millions, of systems companies like Daimler and BMW would
demand if this ABS were successful. Teldix had outlived its usefulness as
an incubator for ABS. Consequently, in 1975 Bosch purchased Teldix's
ABS operation, including its personnel and patents. Teldix's ABS team
was combined with Bosch's own in-house effort.

Bosch was still basking in the glow of success of electronic fuel
injection (EFI), the first electronic component to be widely integrated
into automobiles in all price ranges. When Bosch incorporated the
Teldix ABS project, it placed ABS research development in the same
department as EFI. Engineers who had worked with the reliability issues
of EFI would be helpful in solving similar reliability issues in ABS. Bosch
had to be extremely confident of the reliability of electronic ABS, since
there had been several well-publicized failures with electronic antilock
systems for commercial vehicles in the United States. Any hint of failure
for any reason would risk sullying Bosch's reputation, which was by far
the best in the business. Since Bosch's business rested heavily on automotive electronics, the reliability of ABS put the company's reputation
on the line. By moving the development and production of analog

5 Photo of Bosch ABS-2 system. Reprinted from Manfred Burckhardt, *Fahrwerktechnik: Radschlupf-Regelsysteme* (Würzburg: Vogel Buchverlag, 1993), 277.

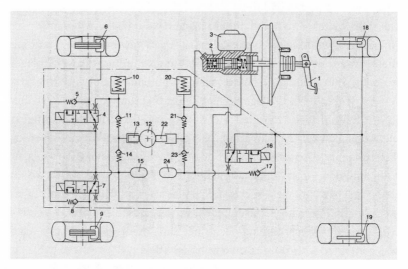

6 Hydraulic plan for the Mercedes-Benz Bosch ABS 2B and/or 2S. Reprinted from Manfred Burckhardt, *Fahrwerktechnik: Radschlupf-Regelsysteme* (Würzburg: Vogel Buchverlag, 1993), 275.

integrated circuits in house, and by contracting American Microsystems to develop and produce digital circuits, Bosch gained control over many of the reliability problems.[43] In addition, Bosch decided to use large-scale integrated circuits in an automobile for the first time, seeking improvement in reliability by reducing the sheer number of parts.[44]

By 1977 Bosch was satisfied with the reliability issue and ABS was available as an option on the 1978 model year Mercedes-Benz S-class sedan (figures 5 and 6). In 1979 the same system was offered on the 1979 BMW 7-series sedan. Between 1978 and 1982 there were 200,000 installations of the Bosch system, meeting the criterion of mass production of an antiskid system for the first time.[45] However, Bosch's antilock system was not yet integrated into a braking system design. Clearly, Bosch's ABS did in fact comprise a system, but it still was not a stand-alone braking system with antiskid capacity designed into it. In 1978 Bosch did not make brake systems; its ABS had to be married to a complete braking system and so was not the first integrated antilock braking system. Consequently, even after Bosch successfully introduced ABS in 1978 and claimed priority for the invention of ABS, another company had the opportunity to trump Bosch.

THE FIRST INTEGRATED ANTILOCK SYSTEM

Antilock systems were developed in different industrial sectors in the United States and Europe. In the United States most antiskid research was undertaken by either automobile companies or brake and tire producers. In Europe R&D led to electronics earlier, and consequently antilock systems were pioneered by electronics companies, including Teldix and Bosch. One important exception to this pattern was the involvement of the largest brake producer in Germany, Alfred Teves, GmbH. Founded in 1906 in Frankfurt-am-Main, Teves was the largest developer and producer of disc brakes for passenger autos, selling 100 million disc brakes by 1978.[46] In 1958 Teves allocated funds to develop an antiskid brake system. Unlike the other systems discussed, Teves wanted to design a new braking system from the start, not an antilock system that could be added on to an existing brake system. Teves committed to electronic control in 1965 but had not made a decision regarding analog or digital. In 1969 Teves debuted an electronic integrated antiskid system at the International Auto Show in Geneva;[47] however, the system was never commercially available. In 1983 Teves introduced another integrated electronic antilock braking system at the International Auto Show. This one, the Teves MKII, did make it to the market as the first integrated antilock system on the 1984 model year Ford Scorpio.

Even a decade before the introduction of integrated antilock braking, engineers at Teves were advocating the benefits of integrating the control system into the braking system (figure 7). To Teves engineers, add-on ABS was more expensive and more complicated. Integrating electronic antiskid control into a brake system would increase its cost, but not by as much as the add-on unit's cost. In addition, Hans Christof

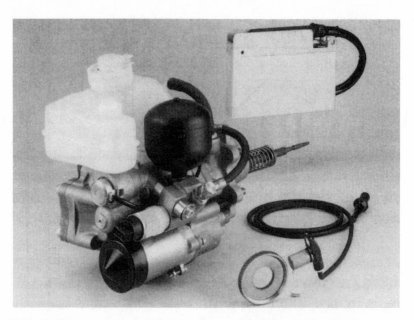

7 Photo of Teves MKII integrated system. Reprinted from Manfred Burckhardt, *Fahrwerktechnik: Radschlupf-Regelsysteme* (Würzburg: Vogel Buchverlag, 1993), 289.

Klein and Werner Fink of Teves argued that a separate antiskid system would be less reliable than one integrated into the braking system.[48] Integrating the antiskid function into the braking system would place the system into a different regime of durability testing, and this would improve not only the reliability of the prototypes but also the reliability of the series production unit. Integrating antiskid into the braking system would elevate the reliability of skid control to the reliability level of a braking system; at least this was the argument put forth by engineers trying to meet this goal. Engineers at Bosch believed that integration of the antilock ability into the braking system threatened to compromise the reliability of the whole braking system and would make safety circuits and failure modes less reliable. While this concern proved to be unfounded, the integrated system did take longer to develop.

MATURE DEVICES AND THE KNOWLEDGE COMMUNITY

In the final analysis, there were several first antiskid devices: Maxaret was the first commercially available system; the Sure-Track was the first production antiskid device; the Sure-Brake was the first four-wheel production antiskid device; Bosch's ABS was the first electronic antilock braking system; and Teves's MKII was the first integrated electronic antilock braking system. These products represent different visions of what constituted antilock braking. As Bucciarelli claims in *Designing Engineers*, the design of a product is a process concerning the questions "What is it?" and "How does it work?" As Schon claims in *The Reflective Practitioner*, a problem cannot be solved until it is defined. These systems look different because their designers were answering different questions. The resources and constraints they had to accommodate were dynamic, changing significantly between 1965 and 1980. And while individual design teams did set their design criteria, as I showed in detail for Chrysler's and Teldix's systems, the knowledge community also played a role in defining the kinds of questions engineers were supposed to focus on.

The period in which Leiber and his colleagues designed ABS at Teldix, Daimler, and Bosch was also one of great economic uncertainties. When Leiber started working on the antilock system in 1964, West Germany was still enjoying a postwar economic recovery; German manufacturing was still escalating in both production capacity and prestige. By 1978 German manufacturing had the prestige necessary for Leiber to show that the Bosch ABS system solved reliability problems, while American manufacturing's reputation for reliability plummeted. The development of the semiconductor industry made it possible for a solid-state electronically controlled ABS to offer reliability rather than inspire skepticism. Teldix and Bosch took advantage of these developments, but certainly cannot claim to have created them. As Oskar Morgenstern once said about economists, design engineers often cannot control precisely the factors they need to in order to make their predictions come true.[49]

Regardless of advertisers' competing accounts of who was first to develop an antilock braking system, the success of ABS shows that investing time and money to meet high performance criteria was the most effective strategy. The success of this strategy may be unique to antiskid devices because of their orientation toward safety, but this detail slipped by the attention of most of the antiskid research projects of the 1960s and 1970s.

The knowledge community around ABS had matured by the 1970s, and the engineers at Teldix, Bosch, and Teves helped steer the community to ask new questions and include new people. As engineers like

Leiber took seriously the vehicle dynamics work done in the 1960s, they also pulled these people more completely into the knowledge community, whereas Mitschke, Milliken and Whitcomb, and the Delft group had been on the margins. The move toward electronic adaptive control, embraced by both the Teldix and Bosch group and Teves, also enlarged the community by adding electrical engineers. But the addition of the electrical engineers and the vehicle dynamicists were not really discrete inclusions. In order for the complex models of vehicle dynamics to be embedded in systems, electronic engineers had to be involved. The design of adaptive circuitry required both the theoretical knowledge of the applied mathematicians and the artifactual knowledge of the electrical engineers. In the end ABS as an artifact bears the knowledge contributions of both.

eight

Public Proprietary Knowledge?

Knowledge Communities between the Private and Public Sectors

I had a curious experience when doing the research for this book. I found the vast majority of the sources I used at the Vaihingen campus of the University of Stuttgart, including many of the non-German-language sources and the Bosch technical reports. I also did research in London at Imperial College, the Public Record Office, and the Institution of Mechanical Engineers, the National Archives in Washington, D.C., the Firestone and Engineering Libraries at Princeton University, Pennsylvania State University, and the Society for Automotive Engineers in Pittsburgh. These are all public libraries and repositories. Other than my interviews with engineers, all of the information that stands behind this account is publicly available. Yet this is a book about a product offered on the market by corporations, who, at least according to conventional wisdom, have every incentive to keep their particular knowledge secret. These corporations required their employees to sign nondisclosure agreements and patent assignment contracts. So why are there over a thousand articles available on antilock devices, all of which were written as the systems were being developed? In fact, it is much more difficult to find after-the-fact accounts; other than this book, I know of only three that have been published: Jean-Pierre Gosselin's in the Swiss publication _Automobile Year_ in 1986, the program for the 1993 Elmer Sperry Award, and a brief article in _Bosch History_ magazine commemorating the twenty-fifth anniversary of ABS. This chapter explores this curious phenomenon: that knowledge we naturally assume to be private is circulated formally and informally in the knowledge community and often in public venues such as conferences and journals. The movement of knowledge is the strongest evidence for the existence and purpose of the knowledge community.

ENGINEERING: A SEMI-AUTONOMOUS PROFESSION?

One of the recurrent themes in the history of engineering is the tension between engineering's status as a profession and the corporate culture in which most engineers work. For well over a century, the majority of graduates from engineering schools have gone on to be employed by companies. Engineers must be responsible to their profession, which often requires sharing knowledge in order to solve socially and technically important problems, but at the same time they must be loyal to their employers, who often require them to sign formal nondisclosure agreements that information be held within the firm. In contrast, other professions, particularly physicians, have found self-employment to be critical to their professional identity; corporate influence can, in fact, delegitimize decisions, a growing problem in America's medical profession as more and more doctors become employees. But for most kinds of engineers, there is simply no living to be made by hanging out a shingle. The work of engineers has always required resources—material, financial, instrumental, and human—available only in corporate or institutional settings. And the need for engineers—that is, the call for what they produce: ideas, designs, and new products and processes—most often comes from the private sector, where companies succeed and fail by their ability to produce and sell the new things engineers design. Consequently, the relationship between corporations and engineers is one of mutual dependence.

However, in looking at the ways engineers actually produce new knowledge, it appears that many engineers share knowledge. Engineers may work for companies, but often it is the engineers who determine the flow of knowledge both into and out of a company. The function of the knowledge community in the development of antilock braking systems illustrates the ways engineers align their professional need to share information with their employers' need to produce unique products. This tension between profession and employment results in an odd oxymoron: public proprietary knowledge. From the perspective of many engineers, the tension isn't as great as it might appear to outsiders, since sharing knowledge involves both offering *and* acquiring knowledge. Therefore participation in the knowledge community can be seen as being in their employers' best interests as well as their own.

Still, the notion of public proprietary knowledge is certainly conflicted. If knowledge is proprietary it belongs to some entity, whereas if it is public it is open to common use. But engineers and their corporate employers have different understandings of what knowledge is publicly accessible. On more than one occasion I have been denied access to certain documents by corporate archivists, only to come across the information in a public location, such as a library or engineering society's conference proceedings. If information is published, how can it be proprietary? If engineers are sharing proprietary knowledge against the wishes and best interests of their employers, how and why do they get away with it? In the case of antilock braking systems, engineers' intellectual needs were at odds with company policy, but the issue was not black and white, and the engineers were not particularly subversive. Instead, a better explanation of the communal dynamic of knowledge production is needed, one which can take into account the dual loyalties of engineers. If we consider only two modes of publicizing new knowledge, patents and publication, it would be easy to think that engineers in the private sector produced only patents while those working at government agencies produced all the publications. However, this simple dichotomy does not correspond to the actual patterns of publication, at least not in the case of antilock braking system development. In trying to tell the story of how an engineering community came together to span company and national lines, one must analyze the ways and the reasons knowledge has been shared, and how engineers and company executives negotiated definitions of what was proprietary and what could be shared, knowing that these definitions might be different. I want to explain how and why engineers in the private sector published so frequently, making allegedly proprietary knowledge available to exactly the people a company should be most worried about: engineers working for competing companies.

I need to backtrack a little before I discuss how the tension between public and proprietary knowledge was mediated in this particular case. It is a bit difficult to recover the mindset of their development since, for the most part, ABS now seems an obvious solution to the obvious problem of skidding. Antilock braking systems, which are ubiquitous on today's cars, allow drivers to maintain steering control of their vehicle even when braking hard on slippery roads, an unquestionable improvement over crashing. But in the 1950s it wasn't clear at all that what drivers needed was some kind of system on their cars to prevent their brakes from locking up. What was clear to these researchers was that skidding caused a significant proportion of accidents, and with the growing volume of traffic, skidding promised to cause more and more people to be injured and even killed in their cars. Given a growing concern over automobile accidents and the escalating number of autos on the road, skidding was seen to constitute a (correctable) public safety problem. But it did not become clear for at least the first decade of research that a better braking system was the only or even the best solution to the skidding problem. Instead, researchers focused on four possible solution areas: roads, tires, automotive systems including braking, and driver behavior. At the Road Research Laboratory they went so far as to consider skidding an interaction problem, overdetermined by these four entangled dimensions. The only way to solve the problem was to break down the interaction problem and make improvements in all of the target areas. Skidding did not appear to have a single solution, and research on all four of these areas does continue to the present day.

While united by their pursuit of a common problem, researchers in all these areas often had little in common; the day-to-day work of road surface engineers, for example, did not have much to do with the practices of engineers trying to improve braking systems. They used different tools and instruments, they thought about different theoretical and mathematical models, and they usually had different sponsors. Each group of engineers working on the skidding problem believed their approach was the most promising; there was no obvious or best solution to skidding at the time. Engineers in each of these groups formed proto–knowledge communities in order to focus on a set of discrete, identifiable, but related problems. When these groups developed into knowledge communities, they acted as the primary organizational unit for engineering R&D.

By the early 1960s an identifiable community of engineers had formed around the questions "Could a better braking system reduce skidding? How?" This group constituted a knowledge community because participants focused on a relatively narrow question, and because they informally controlled access to the community. Access was modulated by accepting certain papers for conferences or journals, but also by limiting a set of desirable solutions to the problem at hand. Michael S. Mahoney calls this an agenda:

The agenda of a field consists of what its practitioners agree ought to be done, a consensus concerning the problems of the field, their order of importance or priority, the means of solving them (the tools of the trade), and perhaps most importantly, what constitutes a solution. Becoming a recognized practitioner means learning the agenda and then helping to carry it out. Knowing what questions to ask is the mark of a full-fledged practitioner, as is the capacity to distinguish between trivial and profound problems; "profound" means moving the agenda forward. One acquires standing in the field by solving the problems with high priority, and especially by doing so in a way that extends or reshapes the agenda, or by posing profound problems. The standing of the field may be measured by its capacity to set its own agenda.[1]

In the case of the nascent antilock community, the agenda formed around a question: How can we design a system that will prevent a car from skidding while braking? But in fact this agenda was more than just a question. From an engineering perspective the question was actually a two-part problem: the system had to first detect imminent skidding, and then react in a way that would prevent skidding. Consequently, the agenda of the knowledge community was at first dominated by problems of measurement. Developing a measurement system that would work reliably in the harsh conditions experienced underneath a car took more than a decade. Antilock systems are research technologies in the sense meant by Terry Shinn, but they are also commodities on their own, with a much broader spectrum of users than scientific instruments.

Tacit agreement on the agenda developed in conferences and collaborations. The first decade of research on skidding prevention focused on these questions of measurement and rarely even mentioned concerns such as mass production, let alone addressing them. Even engineers in the private sector, such as those employed by Dunlop or Ford, were not working in a proprietary mode at this stage. This is not to say that patents were not being filed. In fact, even in this early stage the appearance and use of patents begin to muddy the distinctions between public and proprietary knowledge claims.

Consider Cyril Giles, whose story appears in chapter 3. His was a story about patenting, but Giles worked for the British Road Research

Laboratory, not a private corporation. In some sense this is a peculiarly British story, since it depends on British intellectual property law.[2] This story would not have happened in the United States in the 1960s, since regulations governing the patenting of government-sponsored research made individual researchers unlikely to patent. But Giles's interests in patenting also relate to the question of making public knowledge.

Recall the internal memo Giles wrote to the director of the RRL: "I have recently seen patent specification No. 888,824 of 7 March 1960 for an antiskid braking system using electrical operation. For some time now I have been thinking of systems of this type and the problem of how to get a satisfactory reference voltage when the brakes are applied to all the wheels of the vehicle. Arising from the work on an automatically controlled car, a novel solution to this problem has recently occurred to me."[3] Giles's memo resulted in the pursuit of a patent for the circuit, initially through the Department of Scientific and Industrial Research (which oversaw the Road Research Lab), and by 1963 through the National Research and Development Corporation, a kind of holding company set up by the British government for the commercialization of government laboratory inventions. After internal approval, Giles sent the information to Sefton, Jones, O'dell, and Stephens, a London patent practice.

Giles claimed that his circuit was "a minor improvement at present time, [but] could become very important [if] the antiskid braking systems with electrical operation come into use on road vehicles."[4] He noted that he had seen an article on the Goodyear aircraft antiskid system which could benefit from his design.[5] Giles's research into anti-skid systems for his automatically controlled car clearly led him to examine the leading publications of aircraft design and to engage with the aircraft braking community. Giles served as a vector for knowledge to come into the Road Research Laboratory, but also for knowledge to come into the skid prevention knowledge community. More important, it is clear from his correspondence over this patent that he knew he was contributing to a potentially commercial development leading to anti-lock devices.

His superiors at the RRL were less enthusiastic about his dabbling in aircraft braking systems or in electrical engineering more broadly; Giles worked for the road surface division of the RRL, after all, not some private company specializing in circuit design. But Giles pressed on and angled for a U.S. patent for his circuit. He was not so naïve to think that a market existed for a reference voltage circuit in the United States; rather, he sought to use the patent to open a discussion about the problems of generating reference voltages in road surface characterization testing. He was particularly keen to discuss his experiences with Bendix engineers and believed that a patent would be just the right ice-breaker. In an effort to get the National Research and Development Corporation onboard, he sent a copy of a paper by K. V. Bailey of Bendix to A. F. Cooper, the head of the mechanical engineering group at the NRDC: "I am enclosing a copy of an American paper on an electronic anti-skid brake system for road vehicles which has just appeared and you might find interesting. I think my device would be a distinct improvement on the functioning of equipment of this kind. . . . In view of this American development, you might like to reconsider the decision about patenting the idea in this country only."[6]

Cooper did not budge, and Giles failed to get the circuit patented after two and a half years of negotiation. Given the obstacles to patenting it, Giles chose instead to make the circuit freely available to researchers by writing articles about it.[7] He thus circumvented the patent system when it failed to work the way he wanted it to. Sharing the knowledge of his circuit, one way or another, was paramount. But even in his pursuit of a patent, Giles's priorities were directed toward the communal accumulation of knowledge, not toward self-interested financial gain for his (and the RRL's) invention. Giles sought a patent for the voltage regulation generator not because it was commercially viable, but because patenting it was a way of legitimating it. When that official imprimatur failed to be delivered, he forged ahead with the knowledge community and publicized his new knowledge locally through the meetings that were the ground for the formation of the skid prevention knowledge community prior to about 1965. Giles's story has a very interesting coda. Ironically, his circuit became the basis for voltage regulation circuits in the first generation of ABS. Had he successfully patented it, ABS producers in the United States would have had to license it from him or risk a patent infringement suit.

In contrast to Giles's case, there are several instances in which engineers in the private sector chose not to pursue patents as their initial means of contributing knowledge to the community. Particularly in the early phases of ABS development, through the 1960s, contributing to the development of the engineering community often took priority over developing devices secretly inside the corporation. Engineers thought that they would benefit more directly from sharing knowledge than from keeping it secret. But these benefits came in more than one form: engineers could personally benefit from expanding their own reputation with publications and conference papers, and they also thought the accumulation of knowledge would benefit their ability to produce new designs. An analogy to poker is apt here; an ante of information was required. Once all participants at an informal information-sharing session anted up, a game could begin, with each participant wondering what knowledge the others had and were willing to share. Offering one's knowledge to the group was a gamble: Would the information received justify the knowledge given up? However, not participating in knowledge sharing could have dire consequences, leaving a researcher well behind in the pursuit of a product to market to other engineers. This quandary evidences a moral economy, but it would be naïve to dismiss the gamesmanship of the moment. Regardless of the metaphor, the kind of information a potential community member had and was willing to share played a role in his or her status within the nascent community. It is important to remember that ABS was never envisioned as a product which would be sold directly to consumers; the market for antilock systems was always other engineers and executives at automobile companies. Impressing other engineers thus played an important role in the antilock community, allowing them to develop a market as they developed a product. Engineers at competing companies and automobile manufacturers were treated similarly, since there was enough mobility in the engineers' labor market to allow an engineer to move from a brake or electronics company to an automobile manufacturer. In this case, the free market was not as open as it might seem; buyers and sellers alike participated in the same knowledge community.

To offer evidence of one way an engineer in the private sector used conference and journal publications, I would like to discuss the record of Heinz Leiber. Leiber started his research on ABS in 1964, working for Teldix on the avionics for the F104 Starfighter. In 1958, after being rejected by the U.S. Air Force, it was selected as the nuclear-warhead-capable, supersonic fighter of the West German Air Force. Following Germany's lead, other NATO nations proceeded to license the Starfighter, making it the dominant fighter in the NATO countries, but not in the United States, in the 1960s and 1970s. Thousands of Starfighters were built (and, incidentally, hundreds crashed), so procuring a contract to first produce and then redesign electronics for the plane was significant. Teldix initially produced an electronics package licensed from Lockheed, but by the mid-1960s Teldix was designing new valves for the gyroscopic navigation system, the original design of which proved rather flawed. Leiber was hired to design high-pressure, low-inertia, solenoid (i.e., electrically controlled) hydraulic valves. A low-inertia valve can open and close very quickly, thirty or so times a second. These are small parts, easily held in one's hand. In working on valve development, Leiber became expert in the electronic control of hydraulic parts. By the mid-1960s the Dunlop prototypes on the Jensen, Rolls-Royce, and Jaguar showed that the poor performance of the hydraulic valves was posing as big a problem as the detection of imminent skidding. While the detection problem posed some real theoretical questions and required mathematical models, engineers didn't need a theory of valves; they needed a high-pressure, low-inertia, solenoid valve. They needed what Leiber had. Leiber's valves were patented but, in 1964, unknown to the antilock community.

By attending sessions at the annual meeting of the automotive section of the Verein Deutscher Ingenieure, Leiber became interested in designing an antilock system; he believed, given his experience at Teldix, that he could solve some of the problems on the table. He convinced his superiors at Teldix to create a research team devoted to the design of an antilock system. Given his background in avionics, Leiber entered the antiskid research community somewhat late but as a high-status participant. He clearly came to the table with critical information, which he proceeded to share widely. But Leiber wasn't just a provider of information. Because he had joined antilock research after it was well under way, he had a lot to learn about the dynamics of this community, about its moral economy and the peculiarities of the relationship between theory and artifacts, component design and automobile manufacturing, and about state-of-the-art research techniques in antilock device development. Leiber needed to learn about skidding. His information about valve design provided the community with an incentive to accept him and trade knowledge with him. Starting in 1967 he began to publish a series of articles in both German and English detailing his work on the prevention function (as opposed to the measurement problem) of an

ABS system. In less than a decade he produced about sixty articles on ABS design. Only a few of the details he publicized in his articles were covered by Teldix's patents on the valves.

Leiber's articles were aimed at contributing to the collective knowledge of the ABS community. There was more to gain from this interaction than there was to lose. He believed that Teldix could produce a system whose performance would prove its selling point; this confidence was borne out in the 1969 agreement to collaborate with Daimler-Benz. Leiber was setting different research priorities than other engineers at other companies, devoting his research to a higher performing system, one which would fulfill the twin goals of ABS: it would allow the driver to maintain directional control and it would shorten the stopping distance of the vehicle. In the 1960s no system had even come close to achieving both. For Leiber and the Teldix team, the key to this pursuit lay in the valve: a faster valve would allow the system to maintain higher braking efficiency, which was the key to shortening braking distances. Leiber believed he had this part of the system solved. Therefore he refocused his research on the measurement and control aspects of the system.

He convinced Teldix to invest in electronics at a time when automobile manufacturers were skittish about electronic components. In 1967, at Leiber's request, Teldix hired an electrical engineer, Hans-Jürgen Gerstenmeier, to oversee the design of some electronic control circuitry for Teldix's ABS. Gerstenmeier introduced the concept of adaptive control to ABS development. Adaptive control implies that a system has feedback loops to constantly measure and modify system performance. This was a critical development in ABS research and one in which Teldix took an early lead. In fact, Teldix's adaptive control research influenced publicly funded research on skidding. Research being performed at Delft, Braunschweig, and Cornell Aeronautical Labs was focused on building mathematical models of the dynamics of a braking vehicle. Gerstenmeier's circuitry both benefited and drew from this research. By 1970 Teldix's system design was established and was being published in the *Automobiltechnische Zeitscrift* of the Verein Deutscher Ingenieure.

At this point Leiber and his colleagues started working on prototype production, and they kept publishing in German and English. Back in Heidelberg, the physicist Wolf-Dieter Jonner oversaw the production of dozens of prototypes of ABS, which they tested on frozen Lake Hornavan in Sweden. With this development, Leiber modified his publication strategy, but not to the detriment of his stream of public information. He continued to publish in international automobile engineering journals, but he also began to patent components of the system once he and Jonner had settled on an actual physical design (not just a schematic). Jonner's involvement in this aspect is particularly interesting, since he wrote far fewer articles than Leiber. But Jonner's name appeared on several patents, in both West Germany and the United States. Jonner wasn't working on paper designs, he was working on actual machines, and prototypes were the stuff of patents. Leiber also registered many patents, but not in lieu of writing articles, which often contained pictures of the prototypes. Leiber's, Jonner's, and Gerstenmeier's strategy was to patent the ABS as literally hundreds of individual components. Perhaps unexpectedly, there was no single patent for Bosch's ABS-2. In fact, between the Deutsches Patent- und Markenamt and U.S. Patent and Trademark Office, there are some six hundred patents on ABS components generated by Teldix and Bosch research. This strategy of patenting parts rather than systems had two primary advantages: it allowed Teldix to patent earlier (and more often), which facilitated the flow of new knowledge to the community, and it also made the Teldix system harder to copy, since each element of the system's design was individually protected (and potentially individually licensable). Clearly, intellectual property lawyers and managers were part of the ABS knowledge community. In this case, the public and proprietary dynamics of knowledge production worked in concert.

In addition, Teldix's arrangement with Daimler-Benz involved licensing its design to Daimler. This license was not exclusive, and it did eliminate some potential concerns about finding a market for the ABS system. Most important, it justified the engineers' work to Teldix executives; a contract with Daimler made the project look like it was paying off. One might assume that a contract with Daimler would have made Leiber less interested in publishing, since the marketing aspect of publication became less important with the Daimler contract. But in fact Leiber's rate of publication did not lessen after the Daimler contract. While seeking a larger market for Teldix's eventual product offering helped to justify Leiber's publication strategy, Leiber was genuinely committed to the communal development of ABS knowledge. In fact, the success of Teldix's ABS design team was due to the team's ongoing and central involvement in the ABS community.

Leiber, Jonner, and Gerstenmeier were also personally rewarded for their attention to the community. One of the markers of community development is the emergence of experts. Expertise, in the narrow sense I mean it here, is connected to a knowledge community and is necessarily publicly (but informally) bestowed. Within this structure, the notion of a self-proclaimed expert doesn't make sense. Following Harry Collins's claims in *Changing Order*, the creation of experts is part of the process of research field or community formation.[8] The process of publicly validating experts marks a mature community; the community has to function at a fairly coherent level in order to grant expert status. The informal mark of this status is that others in the community know; if you ask someone who to talk to about a certain problem, you will be sent to the right person. The expert will be widely cited in publications by his or her peers in acknowledgment of his or her status in the community. Two papers that earned widespread citation were Mitschke's and Wiegner's "Blockiervorgang eines Gebremsten Rades" (1970) and Leiber's "Der Elektronische Bremsregler und seine Problematik" (1972). Furthermore, Leiber's, Jonner's, and Gerstenmeier's expert status was ultimately conferred by their 1993 Elmer Sperry Award for Achievements in the Advancement of Transportation. The Sperry Award, granted jointly by the engineering societies representing transportation technologies— planes, trains, automobiles, and ships, that is—credited the Teldix engineers with the work leading to the electronic ABS. Despite the fact that giving credit to three engineers for the work that was done by over a hundred appears to efface the knowledge community, the Sperry Award actually proves, at least in a backhanded way, the existence and importance of the community.

Thus it is the communal aspect of engineering that facilitates the tension between making knowledge public and making knowledge to benefit a single corporate entity. It is, ironically, the benefits of the engineering community which, despite obvious conflict, align the corporation's need to produce unique new products with engineers' social and epistemological needs to share information. Sharing information with an intimate community (these are communities in which everyone knows everyone else) is not at odds with the production of proprietary knowledge; it is part of the process of producing proprietary knowledge.

In a now famous article, "Is Technology Historically Independent of Science? A Study in Statistical Historiography" (1965), Derek de Solla Price focused on differences in the nature of scientific and technological communities and how those communities relate to the nature of the knowledge they produce. He contended that one of the important differences between communities of technologists and communities of scientists is that technologists "want to read, but not write."[9] For Price this phenomenon was due to the largely competitive nature of technological communities. In fact, Price argued, "one might conjecture that the traditional motivation of the technologist is *not* to publish."[10]

From the hundreds of articles produced about antiskid devices and the principles and instruments that underlie them, one can only conclude that Price is partially wrong; engineers do write. Nevertheless, Price's claim that writing is incommensurable with knowledge as private property seems logical. So the question of why engineers do publish provides insights into the tension that clearly does exist between engineers' professional duties and their loyalty to their employers. The case of antilock systems shows that despite working for competing companies, engineers were highly aware of the technological developments made at other companies. They were aware because a structure for sharing that knowledge existed as part of their engineering community. Every nonspecialist automotive conference between 1955 and 1978 presented papers on antiskid devices. From 1963 on there were annual conferences devoted solely to the presentation of papers about antiskid devices. So what motivated these particular engineers to write?

Engineers wrote because they convinced themselves and their companies that publication was the fastest and most effective route to producing more knowledge and therefore better designs. In 1961 an editorial in the monthly magazine of the British Institution of Mechanical Engineers even urged greater participation in conferences, stating, "A wealth of additional information is gained from discussions and correspondence that follow presentation and publication."[11] The informal dynamics of conferences required engineers to present their work in order to be included in ongoing discussions of the field. Engineers who did not share their work were not likely to be included in reciprocal conversations about research at other firms. Presenting one's own work, either in a formal oral or written presentation or informally in conversations about standardizing protocols, research methods, models, or other details, was like an intellectual ante. Participants in the antiskid field acknowledged that 99 percent of the knowledge produced was commonly known by the antiskid community; the 1 percent of information that was held proprietarily for a time was knowledge not yet protected by patents and which might give the lead to another company. The sharing of knowledge was widespread, but also carefully managed.

Patents played an important role in protecting knowledge so that it could be shared. Patents facilitated the exchange of knowledge. This is not to say that engineers frequently read or cited patents in their work; they rarely did. On the other hand, patents did commonly cite seminal journal articles. Patents provided the protection to expose knowledge about devices not yet commercially available. Once a device could be purchased, it could easily be reverse-engineered and many components were readily licensed from one company to another. Patenting played an essential role in the formation of an information economy in the development of antilock systems, but antilock patents themselves were relatively unimportant and, to my knowledge, have never been challenged.[12] Therefore, whether through patents, conference proceedings and presentations, journal articles, or technical reports, writing played an important part in forming and delimiting the antilock knowledge community.

Perhaps the claim that engineers don't write appears true because writing is not the most accessible or logical form through which to transmit engineering knowledge. As many historians of technology have pointed out, engineering ideas are often difficult to communicate in verbal forms. In fact, engineers (like scientists) use many different languages to communicate their ideas and to form and delimit their community. Eugene Ferguson presents an argument for the importance and centrality of visual forms of communication in *Engineering and the Mind's Eye*. Ferguson argues strongly for the proposition that engineering design focuses on the process of communicating ideas (often visual ones) into things. For Ferguson, engineering expression is not scientific, but relies on "intuition, a sense of fitness, and personal preference."[13] Drawing is the primary communication tool, allowing engineers to convey their designs to other engineers as well as the people who will construct these designs. Ferguson explains that "engineering designers convert the visions in their minds to drawings and specifications. In so doing, they solve an ill-defined problem that has no single 'right' answer but has many better or worse solutions."[14] A drawing allows a designer to express an idea to other engineers for collaboration or advice or to technicians who will construct the device. Ferguson's view is that only through explicit visual representation can the form of the object be communicated. Thus illustrations, schematics, and diagrams are also communication tools used by the antiskid community.

Furthermore, as Ferguson implies in the passage quoted above and throughout his book, engineering drawing allows problems to be solved in space, even if not real three-dimensional space. The philosopher Carl Mitcham also sees drawing as a tool for communication, but to a greater extent than Ferguson he emphasizes that drawing in itself produces knowledge. He writes that "drawing is a kind of testing or interrelating of different factors by miniature building. It is not thinking in the sense of conceptualizing or relating concepts; it is thinking as picturing or imagining, and relating specific materials and energies. The designer solves problems of relating parts the way an artist does, by seeing them in practice."[15] For Mitcham, drawing acts as one form of model building, what he refers to as "miniature building." Miniature building is the key labor-saving aspect of design. Engineers can claim time spent on design is labor saved because physical effort is replaced by mental effort; building models consumes less time and money than building prototypes.[16] On an obvious level this is certainly true, but in focusing on visual models both Ferguson and Mitcham fail to appreciate the critical role mathematical models play both as nonverbal (although still textual) forms of communication and as problem-solving tools. Few historians have studied the mathematization of engineering, often burying the issue in debates on school versus shop professionalization or the recurring argument of how scientific theories and engineering or technological theories differ.[17] Yet the widespread use of mathematical tools is one of the few characteristics the different engineering disciplines all have in common with each other.

In addition to the neglect of mathematics specifically, engineering theory generally has held an underexamined position in the history of engineering.[18] A quick survey of the journal *Technology and Culture* shows that the focus of the history of technology is on technology as artifact. Of the nearly three hundred articles published in *Technology and Culture* between 1960 and 1985, John Staudenmeier located forty references to engineering design in his historiographic account, *Technology's Storytellers*.[19] He did not even create a distinct search category for engineering theory. Most of the engineering studies scholars concerned with technological theories have already been mentioned, among them, Vincenti, Layton, Constant, Mahoney, Mitcham, and Ferguson. The examination of the construction of theories by engineers and their role in the culture of engineering has most often focused on a contrast between theory in science and theory in engineering.[20] For the most part, historians and philosophers have obtained their understanding of theory through an examination of scientific theory, so any discussion of theory in engineering naturally gravitates to discussions of how the technological differs from the scientific. If scientific theories are a particular kind of scientific knowledge, then, the argument goes, a parallel relationship should exist between engineering theories and technological knowledge. But all too often the view is presented that the engineer draws theory from science. Too often engineering is seen as the application of scientific knowledge to the problems of society.[21] In *Engineering Design*, Pahl and Beitz take this approach explicitly, writing, "[Designing] is an engineering activity that impinges on nearly every sphere of human life, *relies on the discoveries and laws of science and creates the conditions for applying these laws to the manufacture of useful products.*"[22] However, in the case of antilock systems this contrast is meaningless; even scientists working in the theoretical field of vehicle dynamics admitted that the theories they were building were not science.[23] There are no *laws* of antilock braking systems; the laws governing the operation of the system are contingent on how the system is designed. Furthermore, the goal of these models is predictive rather than explanatory, although philosophers of science have staked out explanation as the sine qua non of scientific theories. While the distinction between predictive and explanatory approaches is not hard and fast, it does denote a different focus in engineering and so-called applied science compared to science.[24] Furthermore, the argument as to whether vehicle dynamics was science or engineering has no historical context; participants simply were not worried about these categories. These categories did have some bearing on their roles in the knowledge community, but not in a predictable or hierarchical way. Knowledge communities are not exclusive, and status comes from a given participant's ability to act as a knowledge vector. The most effective vectors were those who brought valuable information from other communities in which they participated. Leiber's work on avionics and Milliken's and Whitcomb's on aerodynamics and their affiliation with institutions known for work in their areas elevated their status as knowledge vectors, making them multivalent in their ability to bring in key knowledge from other communities.

Peter Galison has used a series of linguistic analogies to describe the construction of multidisciplinary trading zones.[25] I do not offer the notion of knowledge communities as an alternative to trading zones, so much as a different level of analysis; Galison's trading zones apply to a much larger kind of community than I refer to with knowledge communities. Trading zones are not intimate; all the players do not know one another and they do not attend the same meetings or publish in the same journals. Galison posits that pidgins and creoles arise because entrants into the trading zone lack a common language in which to talk to one another. But knowledge communities do not face this problem, except through the usual challenges of human communication and making oneself understood. In fact, a high level of common understanding is assumed by participants in the knowledge community. Recall Mahoney's notion of an agenda, on which participants agree because they share a common goal. The agenda does privilege certain solutions to the problem, and the problem is not defined by incommensurabilities between the knowledge community participants.

In the end I must return to the notion of a knowledge community and its value as a unit of analysis for examining engineering design. Engineering design is difficult to frame using existing epistemology or social structures. It does not make sense to apply rigid epistemological categories to engineering knowledge only to claim that much of what engineers know does not qualify as knowledge; because it fails to explain, it is often situational and not universally true, and it isn't justified in the same ways traditional scientific knowledge is. The category of know-how seems a promising fit, but know-how has been almost completely ignored by epistemologists and philosophers of science. Yet the story of ABS's development is hardly one dominated by unscientific tinkering, though clearly tinkering was important. But know-how doesn't capture the application of complex theories of vehicle dynamics and the embedding of these mathematical models in electronic circuitry that controls precision-machined valves. We need a new set of categories and understandings of what knowledge is needed to describe the design of an artifact. In this account I have focused on a narrative and sequential account of ABS design as a way of analyzing the joint construction of technological knowledge and community. I have focused less on the categories of knowledge that are produced in the design of products, but I adhere to Davis Baird's notion that one kind of knowledge designers produce is the artifact itself. The antilock braking system is clearly an instance of "thing knowledge." The tensions between the complex and rational but often frustrating and surprising processes of making ABS should establish some respect for it as a knowledge form and for its inventors as knowledge producers.

Questions about the social structures of knowledge production also require new framing. The invention of ABS is not a story about disciplines, or institutions. What is notable is the movement of the participants between these better explained organizations. Engineering design emerges at the level of practice, and practices require an intimate level of inspection in order to explain their contingencies. Yet it is difficult to capture the activities of individuals, social arrangements, and knowledge with the tools at hand in science and technology studies. This is evidence of a current tension in science and technology studies between its constituent parts: sociology and anthropology, political science, history, philosophy, and studies of rhetoric and the media. What is lost in this tension is the ability to describe and explain the phenomena in science technology and society. I suspect that the real cost of losing sight of the human dimension is losing the power to understand humanity.

To describe the development of most technologies requires close attention to design, yet this topic is surprisingly understudied. The activities of designing and engineering are so closely related that the verbs *to engineer* and *to design* are nearly interchangeable in common usage. "Henry Ford designed the Model T" denotes an idea similar to "Ford engineered the Model T," in that the product of Ford's work in both phrases is the Model T. In fact, in *Thinking through Technology* Carl Mitcham claims that design is the *essence* of engineering, but still one must analyze the activity of engineering design.[26] What is a design, and what do engineers design? Engineers and nonengineers alike often perceive that designing is the act of inventing and making new products. The designed nature of an automobile, bridge, or jet engine is readily apparent to the most naïve observer; these things were not discovered and they do not naturally occur. Technology is undeniably constructed, and the work of the engineer is intimately associated with the design and construction of new technologies. Furthermore, these developments are purpose-built; engineers create automobiles, bridges, and engines to perform a specific function. However, focusing solely on the *things* engineers design overlooks their more abstract creations. Engineers also construct *ideas*, whether theories, explanations, or mathematical tools and models. Like the bridge or engine, an engineering theory or mathematical tool is designed to perform a certain specific, usually predictive function. Consequently, all products of engineering activity, whether concrete like an automobile or abstract like a mathematical model, emerge from this complicated activity of design. As Mitcham claims, design does form the central activity in the practice of engineering, but design comprises as complicated a process as engineering or even science.

ABS and Risk Compensation

The antilock braking system was invented to improve automobile safety. Whereas many developments in automotive technology have had more complex motivations, ABS was rather simply motivated to prevent wheel locking, skidding, and drivers' losing control of their cars. Certainly the companies that invested in the development of ABS were motivated by profits and the engineers by their own ambitions. But unlike the questions surrounding high-performance engines or gas-guzzling SUVs, the ethical questions raised by ABS are very limited. One might argue that ABS could have adverse effects on a certain segment of the auto-buying public by increasing the cost of vehicles if its use were mandated; the same argument was made in fighting against seat belts and airbags. But in the end these arguments are limited by the fact that these technologies were intended to save lives, without regard for the income of the passengers. In a market economy it must be acknowledged that some people have access to better and more technologies by virtue of their economic resources and that economic stratification is hardly a reason to curtail the production of new expensive technologies when they are aimed at saving lives (and potentially the lives of people who do not own the technology, such as pedestrians). Furthermore, since the introduction of ABS in 1978 its cost has consistently declined and it has become available to more and more car buyers, including those with lower income. As a result, ABS has begun to saturate the market; by the twenty-first century more cars with ABS than without were purchased in the developed world. In 2004 ABS became standard on all new cars sold in the European Union.[1]

Thus, from a market perspective, as well as from the perspective of the engineers' ability to design a product that technically functions as desired, ABS appears to be an unqualified success story. It is an afford-able technological fix to a serious and life-threatening problem; it makes cars safer.

Or does it? By the late 1980s, once ABS saturation began in earnest, studies were undertaken to measure the effects of ABS. The first of these studies was undertaken in Munich by the psychologist K. M. Aschen-brenner. Aschenbrenner and colleagues B. Biehl, B. Wurm, and G. W. Mehr approached the question through three different experiments.[2] In the first, they equipped half a fleet of Munich taxis with ABS and drivers were randomly assigned to vehicles either with ABS or without. The re-searchers recorded 747 accidents during the period of study and found that the involvement of taxis with ABS installed was slightly higher, although not statistically so. Certainly, this study did not provide evi-dence that ABS prevented accidents.

In the second experiment, Aschenbrenner and colleagues installed ac-celerometers (likely ones developed using ABS technology) on twenty taxis, half with ABS and half without. Drivers knew whether the cab was equipped with ABS or not but were not told about the accelerometers. These devices recorded g forces every 10 milliseconds for 3,276 hours of driving. The researchers found that drivers of the ABS-equipped taxis de-celerated quickly more often than those in taxis without ABS equipment.

In the third experiment, trained observers were deployed on the streets of Munich to hail a taxi and request a specific route. Drivers did not know that they were being evaluated. Observers then recorded the quality of driving, including sharp turns, how well drivers held their lane, the following distance to the car in front, whether the driver created "traffic conflicts" in which the driver or another had to take evasive action in order to avoid a collision, and the car's speed. Over 113 trips, drivers of ABS-equipped taxis were consistently observed to take more risks.

This study was widely covered by the media, both in Germany and elsewhere.[3] It spawned a series of further investigations in Canada, France, and Britain, and one by the Organization for Economic Cooper-ation and Development.[4] The Canadian and French studies were test-track studies involving randomly chosen drivers.[5] Both studies found that drivers of cars equipped with ABS, having been told about the ABS and in the Canadian study instructed on its proper use, proceeded to drive at higher speeds with greater pressure on the brake pedal. This was taken as a sign that they drove more aggressively. The higher speed almost exactly compensated for the decline in stopping distance that ABS offered the driver; in other words, stopping distance remained the same as in cars without ABS because the driver's behavior compensated for the technological fix.

Given these results questioning the practical efficacy of ABS, in 1994 the Insurance Institute for Highway Safety sponsored its own studies in the United States. The IIHS started by commissioning its sister organization, the Highway Loss Data Institute (HLDI), to look over claims records and found that ABS-equipped vehicles showed no evidence of reduced claims for accidents.[6] However, the HLDI did not intend to claim that ABS was an ineffective technology, and an advisory published by the IIHS accusing the *Wall Street Journal* of distorting the HLDI study makes the Institute's position clear:

The fact is that antilock brakes can be very effective in helping drivers maintain control during emergency braking on very slippery surfaces. This is their principal advantage. In circumstances in which a vehicle with regular brakes may skid out of control, antilocks can help prevent the skid and allow the driver to steer away from a possible collision. But these circumstances are rare for the average motorist, which is probably why HLDI found no reduction in claim frequencies for cars with antilocks.

Because antilocks have been promoted as a wonder technology enabling motorists to stop on a dime and thus avoid collisions, expectations for them have been unrealistic. Antilocks do *not* offer substantially improved stopping distances on most road surfaces. Only on sheer ice or otherwise very slippery surfaces are there pronounced improvements for cars with antilock brakes and, on some surfaces such as gravel or loosely packed snow, antilocks can actually *increase* stopping distances.[7]

By 1996 the overall performance of antilock systems had become even harder to assess. Explanations were sought for the anomalous results, leading to a pitched argument about the rigor and validity of risk compensation theories. Risk compensation was not a new idea in the 1990s, although studies involving the effects of air bags and antilock systems created a whole array of new empirical evidence. Risk compensation is the idea that individuals have an internalized threshold for risky behaviors, and if a behavior one engages in becomes safer due to technological reasons, one will tend to increase one's exposure to risk in other ways. For example, skiers and cyclists who wear helmets take more risks than those who don't wear helmets, in part because they feel safer and are subconsciously correcting for their exposure to risk. This theory is used to explain why safety devices fail to produce their desired social effects; for example, seat belts proved to save fewer lives initially than was expected in the 1970s.[8] It is not an argument that the devices do not work, but rather that humans compensate and maintain a constant level of risk in their daily activities.

The most vocal proponent of risk compensation in the 1990s was Gerald Wilde, a psychologist from Queens University in Ontario. Wilde picked up on the ABS controversy with the Munich taxicab study, using it as a key case study in his book on risk compensation, *Target Risk*, published in 1994. He also described or cited many of the other ABS studies of the early 1990s. Wilde's notion of risk homeostasis explained the subconscious way humans calculate risk to maintain a constant level of risk.[9] Discussions of risk compensation with ABS also made it into the mainstream media, in the *Wall Street Journal*, for example.

In fact, it was specifically the risk compensation argument that prompted the IIHS's advisory in 1994. As is clear from the IIHS's reaction to the article, not all of the participants in the debate about ABS's efficacy were swayed by the risk compensation argument. The IIHS advisory called risk compensation discredited:

Ms. Miller [*Wall Street Journal* reporter] resurrects the discredited risk-compensation hypothesis. According to this hypothesis, people adjust their behavior and take more risks behind the wheel to offset the car safety improvements mandated by federal standards. The risk-taking hypothesis has been repeatedly discredited. Overwhelming evidence in the scientific literature shows that mandated safety features—as well as features now being offered in response to consumer demand—have substantially reduced crash deaths and serious injuries. There isn't a shred of empirical evidence to support the hypothesis that drivers take more risks when operating cars with features that reduce the likelihood of injury or death in a crash.[10]

Both the anomalous results and the irritation over the risk compensation or homeostasis explanation led to a more extensive study by the IIHS, which sought to break down the ABS accidents into categories in

Table 2 *Accident Categories for Vehicles*

Who is killed	Crash type	Road surface condition	GM cars (1992 with ABS vs. 1991 w/o ABS)	Other vehicles (with ABS vs. w/o ABS)
All	*All*	All	1.03	1.16
		Wet	0.92	1.18
		Dry	1.06	1.15
	Single vehicle	All	1.17	*1.28*
		Wet	0.99	*1.26*
		Dry	*1.21*	1.28
	Multiple vehicles	All	0.95	1.07
		Wet	0.89	1.13
		Dry	0.98	1.05
People in ABS-equipped vehicle	*All*	All	*1.24*	1.26
		Wet	1.08	*1.40*
		Dry	*1.30*	1.23
	Single vehicle	All	*1.39*	1.45
		Wet	1.19	*1.65*
		Dry	*1.44*	1.42
	Multiple vehicles	All	1.13	1.06
		Wet	1.02	*1.20*
		Dry	1.19	1.02
People in non-ABS vehicle	*Multiple vehicles*	All	0.80	1.01
		Wet	0.87	0.91
		Dry	0.78	1.03

Source: Figures from Insurance Institute for Highway Safety, "Antilock Brakes Don't Reduce Fatal Crashes; People in Cars with Antilocks at Greater Risk—But Unclear Why" (news release).
Note: A ratio greater than 1.0 means vehicles with antilock brakes were more likely to be involved. Numbers in italic show a 20 percent or greater accident rate for ABS-equipped vehicles.

order to find cause. The study produced the statistics in table 2. The headline of the news release announcing the findings was "Antilock Brakes Don't Reduce Fatal Crashes; People in Cars with Antilocks at Greater Risk—But Unclear Why." In table 2, any result greater than 1.0 indicates a situation in which antilock-equipped vehicles were *more* likely to be involved in accidents. Most notable is the singling out of a particular kind of crash that seemed to be significantly more common in antilock-equipped vehicles: fatal crashes of single cars. These accidents are often referred to as "run-off-the-road" accidents, and they have a number of proximate causes, including drivers falling asleep, bad road conditions, and animals in the road. But even here the results were not clear; for example, GM cars with standard antilock systems actually saw an overall reduction in fatalities.

Three explanations for these results emerged: the IIHS criticized drivers for risky behavior and for not properly using the systems; the risk homeostasis group predictably cited risk compensation as the cause; others, including engineers, cited social explanations, such as that more parents bought cars with safety features for their children, who were inherently less experienced and more risky drivers. The first two explanations generated a small cottage industry of social science and public policy scholarship. Another explanation argued that in an accident drivers would panic and oversteer, thus running their car off the road. This didn't happen with conventional braking, since any steering failed to produce a change in the direction of the vehicle.

With the support of the IIHS, a group of antilock system producers, including Bosch, Teves, Kelsey-Hayes, and Bendix, founded the ABS Education Alliance. This nonprofit organization was dedicated to educating drivers on how to use antilock brakes properly. The ABS Education Alliance set up a website (www.abs-education.org) offering information for drivers, driver education teachers, state motor vehicle agencies, and others interested in retraining drivers to use antilock systems, specifically to stop pumping their brakes during panic stops. Drivers are encouraged to find a safe place where they can test the system and become comfortable with the sensations it produces. This is important because some drivers report that when they felt the vibration of the antilock system they backed off the brakes. On cheaper systems, the vibration transmitted to the driver can be significant, so knowing it is normal and not a malfunction is an important dimension of retraining drivers.

What was most interesting about the ABS education movement was the extent to which it was a direct challenge to one of the sacred concerns during the design of antilock systems. Beginning in the late 1950s the antilock knowledge community agreed that in order for antilock systems to have any possibility of effectiveness, they could not require any new training of the driver. In fact, the usual understanding was that antilock systems had to be invisible to the driver (i.e., undetectable in both operation and user's habits). Yet in 1996 Brian O'Neill, the president of IIHS, claimed, "It's a good idea to try antilocks out. When it's wet, go somewhere off road like a parking lot and practice hard braking so the antilock feature is engaged. See how the brakes feel because it's important to 'unlearn' past braking habits and keep hard continuous brake pressure instead."[11]

The U.S. agency for road safety (formed in the wake of the Nader-Ribicoff hearings of the late 1960s, which questioned car manufacturers about their commitment to safety), the National Highway Transportation Safety Administration, under the oversight of Elizabeth Mazzae and Garrick Forkenbrock, began researching ABS and established the Vehicle Research Test Center. The agency's main concern was single-car accidents where drivers drove off the road, so-called run-off-the-road accidents.[12] Researchers were willing to entertain several explanations: single-car accidents were a statistical anomaly not requiring attention; what the NHTSA termed "early models" of ABS might be flawed and new algorithms already embedded in the newer systems of the 1990s would correct this; ABS equipment could have negative effects in certain situations, extending stopping distances; drivers were using ABS improperly; and risk compensation behavior. To determine which of these had the best empirical basis, the NHTSA commissioned seven (which by the end of the project had become nine) different studies with four goals:

1. Determine whether newer models continue the pattern of reducing multiple car accidents while increasing run-off-the-road accidents.
2. Determine why the number of single-car accidents increased.
3. Determine whether ABS on light trucks shows the same pattern of increase.
4. Develop a consensus explanation between NHTSA, the automobile and insurance industries, and other stakeholders as to the findings.

The NHTSA completed the last of the reports from these studies in May 2003. The reports were made publicly available, and many were also presented at the meeting of the Society of Automotive Engineers in 1999. What is notable about the reports is the extent to which they fail to shed light on the problems they set out to illuminate. They did find that the anomalies lessened with newer models of ABS, which may add credence to the claim that improved algorithms in fifth- and sixth-generation ABS were able to improve performance.[13] The NHTSA studies found little evidence for risk compensation behavior among the drivers they tested and surveyed. They did find that many drivers were unaware of proper use of ABS, and thus their conclusions were used by the IIHS and ABS Education Alliance to support further educational programs. But in terms of truly explaining the ABS anomaly, the NHTSA did scarcely better than any other agency, despite its resources. The NHTSA created a website for drivers and consumers explaining its findings as well as how to properly use an ABS.[14] Transport Canada followed the U.S. agency's lead and created a website and programs for driver education.[15]

In the twenty-first century there have been several more international studies on the efficacy of ABS and programs to better educate drivers about ABS use.[16] Obviously the problem is built into the scenario in which one uses ABS. This is a device designed to work only under the exceptional circumstances of the panicked maneuver; therefore, practicing to use the system constitutes a problem in two ways. First, it is inherently difficult to practice panicking; second, it is unclear whether practice will really affect one's behavior in a panic situation. Regardless of the education programs, one troubling result remains: evidence that ABS actually reduces crashes is difficult to find.[17]

Most striking in the history of ABS is the degree to which a fundamental understanding about the users of ABS was both correct and misleading. Designers of ABS in the 1960s and 1970s assumed that drivers could not be asked to modify their driving habits; that assumption became the basis of one of the design's criteria. In the end, it seems that ABS does require drivers to change some of their habits, but it also seems that the engineers were right about the inability of drivers to change.

If there is a lesson to be learned from this case and its long-term design outcome it is about the role of the human user. For many engineering design projects users are more often thought of as potential misusers, which leads to engineers' inherent frustration with users. As a result the inadequacies of the user are easy to dismiss; engineers believe they should concentrate on designing a good system and let common sense sort out the users. Anecdotal evidence from individuals who believe that ABS saved their life is plentiful. Remember Starrion from the beginning of this book? Yet in the case of ABS, the inability of users to accommodate their driving habits threatens to undo the promise of improved public safety that motivated its invention and production. If ABS fails to save lives, then what is its purpose?

The other dimension of concerns about the effectiveness of ABS that is interesting in the context of this study is the way that social science and public policy researchers interested in the question quickly formed their own knowledge community. An interdisciplinary group of psychologists, statisticians, policymakers, engineers, and others came together after the Munich taxicab study to investigate the question of whether antilock systems produced their anticipated social benefits. The ambiguous results breathed life into this knowledge community, which used similar tactics in coming together at specialized conferences, establishing an agenda, publishing their research in a few selected journals (for example, note the number of articles in the journal *Accident Analysis and Prevention*), and bringing in new members with new special skills. Even the somewhat nasty disagreement about the viability of the risk compensation argument has not destroyed this knowledge community.

These two observations come together in an interesting paper published in the IEEE *Transactions on Systems, Man and Cybernetics—Part A (Systems and Humans)* in 2004. Two mechanical engineering professors, Shinsuk Park and Thomas B. Sheridan, wrote "Enhanced Human Machine Interface in Braking," a study on the human kinematics of brake pedal use. They studied the motion of the human leg while driving and proposed a new system which would accommodate the driver's leg action and provide the driver with kinesthetic feedback about braking and driving conditions. Here the user is being taken seriously into account in the mechanical design. And yet it seems ironic that the solution presented is yet another technological fix, liable to use and misuse by human users.

Notes

---

1. DESIGN AND THE KNOWLEDGE COMMUNITY

1. "This Car Saved My Life," http://www.epinions.com.
2. The acronym ABS, "Anti-Blockiersystem," is a trademark of Daimler-Benz used to refer to the system developed by Teldix and manufactured and marketed by Bosch. The acronym, but not the German phrase, is trademarked. However, in the introduction I use the phrase more generally, as is common practice in English.
3. Friction coefficients have to do with the physics of friction and the material properties of the two surfaces in contact; for the most part they are experimentally determined. There are exceptions to the rule that sliding tires encounter lower resistance, such as deep sand or snow, where the snowplow action of the wheel means that the whole wheel encounters more resistance, though not through frictional properties.
4. See Pickering, *Science as Practice and Culture*, 3. Pickering defines culture and practice in the following way: "Culture denotes the field of resources that scientists draw upon in their work, and practice refers to the acts of making [and unmaking] that they perform in that field" (n. 1).
5. In *Thing Knowledge: A Philosophy of Scientific Instruments* Davis Baird also makes a similar claim for the importance of artifacts in the production of and as bearers of scientific knowledge.
6. Like most historians of technology I find the notion of technology or engineering as applied science to be problematic, as it tends to imply that engineering does not make its own knowledge but simply applies scientific knowledge. See Layton's "Mirror-Image Twins." For the history of the notion of applied science in the American case, see Kline, "Construing 'Technology' as 'Applied Science.'"
7. In the case of knowledge communities, I use the word *membership* to mean informal participation. However, knowledge communities are so small that individuals know each others' roles and specialties, so there are (again, informal) standards for informal participation. If one's involvement falls below some (again, informal) threshold, the community tends to forget about one's participation; one really ceases to be a resource to the community. In the period I am discussing in this book, participation tends to mean face-to-face interaction at meetings as well as in the workplace, though electronic communication has changed that in recent years.
8. Vincenti, *What Engineers Know*, 232.

9. Of course this is a vast oversimplification, but still a common enough opening claim in epistemology textbooks. Most modifications of the definition of justified true belief involve adding criteria rather than arguing that knowledge is not justified true belief.

10. The development of philosophy of experiment since the 1980s provides a welcome expansion of the notion of knowledge in philosophy of science, yet much of that literature still focuses on theories and sees experiments as theory-laden. See Radder, *The Philosophy of Scientific Experimentation*.

11. Baird, *Thing Knowledge*, 12.

12. Morison, *From Know-How to Nowhere*, provides an account of the dual mental and manual aspects of engineering and craft know-how. It is a mistake to collapse the two dimensions of know-how into the notion of skill, since not all know-how is embodied.

13. Bucciarelli, *Designing Engineers*, 1–2.

14. Ibid., 9.

15. Ferguson, *Engineering and the Mind's Eye*, 2.

16. One is reminded of Edison's oft-quoted line that invention is 1 percent inspiration and 99 percent perspiration. In the corporate case, inspiration and perspiration, as Edison referred to them, occupy two different hierarchies: the executive and engineering branches of the corporation.

17. A philosophical research program at Delft Technical University in the Netherlands focuses on the "dual nature of technical artifacts," that is, the relationship between their physical instantiation and their operational function. See Meijers and Kroes, *The Empirical Turn in the Philosophy of Technology* and the Delft philosophy of technology group's manifesto at http://www.tbm.tudelft.nl.

18. I grant that the design of the body and therefore the aesthetics of an automobile are often easily separated from the design of its mechanical systems. But these are two different projects; the design of the body and the design of the engine require two sets of constraints and two groups of researchers. This is not the issue to which Bucciarelli refers. Using Galileo's cantilever beam as an example, he argues that the questions "What is it?" and "How does it work?" are inseparable; they constitute an epistemological problem rather than an argument about the relationship of form and function.

19. Pahl and Beitz, *Engineering Design*, 17.

20. Ferguson, *Engineering and the Mind's Eye*, 37.

21. The closer a design gets to actual production design, the more discrete the analysis phase may become. In the final phases analysis can be quite separate from conception, but this does not hold for their earlier stages of development.

22. Thinking back to Baird, it is worth noting that this common model of scientific method privileges theories as the ultimate knowledge product of scientific activity. Thus the common representation of the scientific method, as found in both introductory textbooks, likely ill describes many scientific

and engineering activities, which aim at making artifacts. If we consider the synthesis of molecules a form of artifact production, then this distinction puts much of chemistry into the artifact-privileging rather than theory-privileging category.

23. Bucciarelli, *Designing Engineers*, 3.

24. Understanding conception and function independently is a particularly problematic issue when dealing with systems. The system itself is far greater and more complex than its parts. The work of Thomas P. Hughes focuses on the complexity of systems and the interdependent aspects of their conception, operation, and function. See especially *Networks of Power*.

25. Ooudshorn and Pinch, *How Users Matter*.

26. Ferguson, *Engineering and the Mind's Eye*, 32.

27. Vincenti, *What Engineers Know*, 209.

28. Polanyi, *Personal Knowledge*, 328.

29. Akera, *Calculating a Natural World*, 13–16. Akera's introduction focuses on the problem of metonymic devices, that is, a part that stands in for a whole, which is a rather elegant way of stating this problem with case studies and generalizable knowledge drawn from case studies.

30. See, for example, the International Technology Roadmap for Semiconductors as an example of such a technological planning and forecasting document, http://www.itrs.net.

31. Schon, *The Reflective Practitioner*, 94–95.

32. See Rammert, "Two Styles of Knowing and Knowledge Regimes."

33. See particularly Petroski, *Design Paradigms* and *To Engineer Is Human*.

34. Vincenti and Rosenberg, *Britannia Bridge*, 68.

35. Pahl and Beitz, *Engineering Design*, 18.

36. *Closure* is a notion developed in Bijker's *Of Bicycles, Bakelites and Bulbs*, meaning the social and technical processes through which artifacts adopt their finalized form.

37. Joerges and Shinn, "A Fresh Look at Instrumentation," 2.

38. Petroski, *Design Paradigms* and *To Engineer Is Human*.

39. Kidder, *Soul of a New Machine*, 30–31.

40. I do not mean to imply that internal and external social influences are completely separate. I simply want to emphasize the importance of social dynamics within the company as well as social influences emanating from other places. Neither is inherently more important than the other.

41. *Elmer A. Sperry Award 1993* (program).

42. Collins, *Changing Order*, chapter 6.

43. It is worth noting that this promise often went unfulfilled. While it worked for Leiber, Jonner, Gerstenmeier, and Bosch, leading designers at other companies were less successful at parlaying their high status in the knowledge community into the market success of the devices they shepherded to the market. Because

so many nontechnical factors go into market success, the implication that a simple one-to-one relationship exists between good design and market success is simply naïve, but this isn't to say that the belief isn't widely held among design engineers and even CEOs.

44. Baird, *Thing Knowledge*, chapter 4.

45. Most engineers felt they had a stake in creating a common set of measurement standards and that measurements, mathematical models, and protocols should all be shared.

46. The most famous testing site for antilock brakes is in Arjeplog in northern Sweden since large frozen lakes are a great resource for testing antilock systems.

2. GENEALOGY OF KNOWLEDGE COMMUNITIES

1. Baird, *Thing Knowledge*, 4.

2. See Braun and Macdonald, *A Revolution in Miniature*, for a discussion of the resistance to electronics by the automotive industry. Further complicating the introduction of electronic antilock systems were patterns of innovation in the automobile industry itself. Improvements in automotive technology until the 1970s (that is, until the so-called electronics revolution, of which ABS was the first shot) tended to move in fits and starts rather than in slow and gradual improvements. Consider the introduction of disc brakes, automatic transmissions, emissions control, or fuel injection; all of these represent discontinuous shifts from the older technologies they replaced. The antilock braking system represents a change of a different nature, in which the continuities with older brake systems are much greater. The antilock system was designed to be added on to an existing braking system, not to replace it altogether.

3. Hobsbawm, "Introduction," 1; Hobsbawm, "Mass Producing Traditions," 263.

4. Mom, "An Improved Safety Device"; Robert Bosch, GmbH, "Vorrichtung zum Verhüten des Festbremsens der Räder eines Kraftfahrzeuges."

5. Just such a claim is made in *Elmer A. Sperry Award 1993*.

6. Moynihan, "Epidemic on the Highways," 16.

7. Goodhart, "Statistics and Road Accidents," 433.

8. S. C. Davis, *Transportation Energy Data Book*, 11–12.

9. For the United States, see Albert, "Order out of Chaos."

10. Bell, "The Motor Vehicle in Relation to Accidents and Injuries," 461.

11. Jameson Wetmore calls this the "crash avoidance paradigm," which ran through the early 1960s, when it was replaced with a different way of thinking about road safety, which he calls the "crashworthiness paradigm." His argument is compelling as an expression of public policy thinking about automotive safety, yet ABS seems to be out of phase with these paradigms, since it is clearly a crash avoidance technology introduced in the late 1960s and 1970s. See Wetmore, "Systems of Restraint," 31–32.

12. Chayne, "Automotive Design Contributions to Highway Safety," 73.

13. For a similarly debated case involving the political dimensions of introducing automobile airbags, see Wetmore, "Redefining Risks and Redistributing Responsibilities."

14. Bradley and Wood, "Some Experiments," 50.

15. Pyatt, *The National Physical Laboratory*, 119.

16. *Research on Road Safety.*

17. Self, "Breaking Skids Betters Braking"; Bent, "No More Skids for Airplanes!"

18. A letter from G. N. Eisenhart on 13 February 1944 details the supply problems of Fighter Squadron 32 caused by blowouts from aircraft skidding on carrier decks. Eisenhart claims that the problem was due to the fact that the wheel, tire, brake, and shock were all packaged as a unit by Goodyear, making the replacement of a tire a very complicated and expensive undertaking, especially in the South Pacific. Bureau of Aeronautics Confidential Correspondence F13–2(2), U.S. National Archives.

19. Keyser, "Electrical Prerotation of Landing Gear Wheels," 454.

20. Glanville, "Summary," ix–x.

21. The Motor Industries Research Association (MIRA) was one of several collective research associations set up after World War I. The British government created the research associations as sites for collaborative research. Though some government funding aided MIRA, it still pulled in a large percentage of its operating budget as contracted R&D. This can be contrasted to the situation in the United States, where the National Science Foundation focused on funding "basic research" at the expense of applied; applied research was largely the provenance of the military, and industry collective research was often discouraged through more restrictive antitrust laws. It is interesting to see the reappearance of such commercialized (as opposed to militarized) cooperation between industry, academia, and the government on the post–cold war R&D landscape.

22. Kinchin, "Disc Brake Development," 203.

23. Lister and Kemp, "Skid Prevention," 391.

24. Rixmann, "Neues aus der industrie," 106; Leiber and Limpert, "Der Elektronische Bremsregler," 181.

25. Strien, "Trends."

26. *Bremsen-Handbuch*, 14–16.

27. J.-P. Gosselin, "How ABS Was Born," 62.

28. Ibid.

29. Kullberg, Nordström, and Palmkvist, "Antilock Braking System"; Nordström, "Provning av antilåssystem."

30. Arjeplog Test Center in Swedish Lapland, "Milestones," Arjeplog Test Center, www.arjeplogtestcenter.com.

31. "Ice Tracks as Far as the Eye Can See," *Arjeplog Times,* www.arjeplogtimes.com;

Karl Hardell, "Swedish Car Test Region," paper presented at Rural Clusters 2007: Conference on Cluster Development in Small and Remote Communities, www.ruralclusters2007.com.

32. Managers and executives may have had a different opinion on this question.

33. The period 1965 to 1975 was one of considerable inflation in Britain, making the price differential all the more sharp.

34. Kuhn, "Second Thoughts," 297.

35. Constant, *Origins of the Turbojet Revolution*, 10.

36. This points to another difference between the effect of a revolution in science and one in technology: in technology the losing paradigm may still produce a useful artifact, whereas the newly dominant scientific paradigm will completely drive the old one out of the discipline.

37. Dosi, "Technological Paradigms," 148. This is also one of Kuhn's main points.

38. Ibid., 152. See also Schon, *The Reflective Practitioner*, 94.

3. THE BRITISH ROAD RESEARCH LABORATORY

1. Shelburne, introduction, viii.

2. Ibid., vii.

3. See *Road Research Laboratory Technical Report* 122, 1952.

4. "Report of Subcommittee B," xxvi.

5. Ibid., xxix.

6. Pyatt, *The National Physical Laboratory*, 119.

7. Bradley and Wood, "Some Experiments."

8. *Research on Road Safety*.

9. Giles and Sabey, "Skidding as a Factor in Accidents," 28–30. See also *Research on Road Safety*; Lister and Starks, "Experimental Investigations"; Giles and Grime, "The Skid-Resisting Properties of Roads and Tyres"; Jones, "The Skidding Behaviour of Motor Vehicles"; Kummer and Meyer, "Verbesserter Kraftschluß."

10. One significant advantage the RRL had was its access to nationally standardized police reports of automobile accidents. This facilitated the compilation of national road statistics and led the RRL to focus much earlier on skidding as a preventable problem. Both the Federal Republic of Germany and the United States kept accident statistics on a state (or Land) basis, and they did not standardize the police accident reports from one region to the next. As a result, national figures were hard to arrive at, and skidding as a national problem was less obvious. In the United States, Virginia was reputed to keep the most effective statistics; see Mills and Shelton, "Virginia Accident Information Relating to Skidding," 9, 11. In Germany the most elaborate statistics were kept by the Land of Schleswig-Holstein; see Wehner, "Accidents Involving Slippery Road Conditions in Germany," 45. Note that neither Virginia nor Schleswig-Holstein represented the most densely populated regions of their respective nations.

11. For Swedish statistics, see Kullberg, Nordström, and Palmkvist, "Antilock Braking System for Passenger Cars."

12. Coulomb's law, which states that the coefficient of friction does not change with velocity, is invalid for wet surfaces because the water on the road acts as a lubricant. Water also forces the oils on the road surface to rise to the surface, making a wet road literally greased.

13. For a given braking efficiency, stopping distance is strictly a function of the kinetic energy of the vehicle.

14. Normann, "Today's Speeds," 5; Normann, "Braking Distances of Vehicles," 159.

15. Vincenti, *What Engineers Know*, 9.

16. For what is known of friction in the 1950s, see Palmer, "Friction," 55; Barwell, "Friction and Its Measurement," 141.

17. When the Department of Scientific and Industrial Research was created in 1916, one of its goals was to facilitate communication between research projects in different departments and industries. As much as it was a funding agency, the DSIR was also imagined as an information distribution agency.

18. Discussion of Giles and Grime, "The Skid-Resisting Properties of Roads and Tyres," 50.

19. Baker, "Causes of Skidding Accidents," 26.

20. Giles and Sabey, "Skidding as a Factor on the Roads of Great Britain," 30.

21. Ibid., 44.

22. The quotation used as a heading is taken from Forbes, "Driver Knowledge," 57.

23. Ibid., 51–58. See also Fiala, "Die Wechselwirkung zwischen Fahrzeug und Fahrer," 222; Oppenheimer, "Safer Braking Systems," 90; Oetzel, "When Are Brakes Adequate?," 243; Halsey, "Design Cars for Poor Drivers," 17.

24. Discussion of Giles and Grime, "The Skid-Resisting Properties of Roads and Tyres," 48.

25. Lister, "Brake Performance Measurement," 49; Lister and Starks, "Experimental Investigations," 31; Lister and Stevenson, "Fifth Wheel for Measuring Speed and Braking Distance."

26. In RRL tests, locked wheel braking consistently gave better stopping distances, despite the fact that locking the wheels actually decreases friction and therefore should lengthen braking distance. The reason for this is practical. If a car could maintain its brakes at a point of impending skid, that is, just before the wheels are about to lock, its braking distance would be shorter. However, to prevent skidding the car cannot actually maintain impending skid; one wheel will lock up because of irregularities in the brake lining or the road surface or imbalances in the brake pressure. One of the keys to antiskid devices is that they keep braking efficiency at the impending skid level in order to maintain stability as well as shorten braking distance.

27. Collier, "Two Methods of Aircraft Skid Control"; Bent, "No More Skids for Airplanes!," 63; Bent, "Automatic Brake Control," 140–42; "Stopping without

Skidding," 504; Self, "Breaking Skids Betters Braking," 49–53; Gunsaulus, "Rolling Wheels Gather No Skids," 189–95.

28. Despite this fact the installation of antiskid units on aircraft always included a skidding indicator light on the instrument panel. Its function must have been more for system check than for the pilot to react to.

29. Keyser, "Electrical Prerotation of Landing Gear Wheels," 454–59.

30. "Automatic Wheel Landing Wheel Braking," 461; "Anti-Skid Device," 248–54; "Wheel Lock Inhibited Braking," 940; Lister and Kemp, "Skid Prevention," 382–91; Kinchin, "Disc Brake Development," 203–11; *Research on Road Safety*, 414–19.

31. I discuss braking distribution issues in chapter 4. Braking distribution problems exist only when a single master cylinder must control the operation of multiple brakes, as is the case on most automobiles, but not on this particular experimental setup.

32. Randle, "From Sidecars to Silicon"; Kinchin. "Disc Brake Development."

33. On devices which engaged at a certain rate of vehicle deceleration (as opposed to wheel deceleration), cornering could pose a problem, since it generated g forces and, in a few cases, erroneously triggered the antilock systems.

34. Ford engineers also encountered this problem in their work with Kelsey-Hayes's antiskid device. See Madison and Riordan, "Evolution of the Sure-Track Brake System."

35. It is worth noting that the cost of the device as a percentage of the cost of the whole vehicle is more critical than the absolute cost. When the Maxaret appeared commercially on a Rolls-Royce Silver Shadow in 1966, it accounted for about 10 to 15 percent of the vehicle's cost. Given a £500 Volkswagen Beetle, the Maxaret would have more than doubled the car's cost. These financial relationships ruled it out of the market completely for all but novel applications. Its percentage in the total cost of an airplane, or even in a tractor, was much more acceptable, around 10 percent. The wide acceptance of ABS in the 1990s is in significant part due to its cost being driven below $1,000 and the average cost of a vehicle being more than $15,000. The 10 percent cost threshold has remained surprisingly constant since the 1960s for this and other safety devices.

36. "Sticky" rubber has been extensively researched by the tire and rubber industries. Rubber compounds have been developed that are highly effective at increasing traction, but they also significantly lower fuel economy and wear out quickly. Most of the development of sticky rubber has been with an eye to Formula One racing cars, where cornering and braking are priorities over fuel economy and tread life.

37. Hysteresis is the study of the way materials permanently change under load.

38. Giles to McIvor, 12 March 1962, Public Record Office (hereafter PRO), Department of Scientific and Insustrial Research (hereafter DSIR) 28/406, UK Patent Application 31986/63.

39. Ibid.

40. Ibid.

41. Howe to McIvor, 13 April 1962, PRO, DSIR 28/406.

42. Giles to McIvor, 18 April 1962, PRO, DSIR 28/406.

43. Communications between Giles and Howe make it clear that Howe did not see the tests. There is no communication between Giles and Lister in the Public Record Office files of the DSIR pertaining to this patent.

44. Meade to McIvor, 13 June 1962, PRO, DSIR 28/406.

45. Internal memorandum: Meeting notes of Meade and Giles, 8 August 1962, PRO, DSIR 28/406.

46. NRDC Patent Questionaire, no date, PRO, DSIR 28/406.

47. Ibid.

48. Lister to Cooper, 19 February 1964, PRO, DSIR 28/406.

49. Giles to Cooper, 14 October 1964, PRO, DSIR 28/406.

50. Sefton, Jones, O'dell, and Stephens to DSIR, 16 November 1964, PRO, DSIR 28/406.

51. See Mowery and Rosenberg, *Technology and the Pursuit of Economic Growth*, 98–119; Hill, *Co-operative Research in Industry*; P. S. Johnson, *Co-operative Research in Industry*; Dodgson, *Technological Collaboration in Industry*.

52. The impetus for research associations came from the same sources that pushed for the creation of the DSIR in 1916. Research associations were industry research collectives initially funded by Parliament with a £1 million grant. The research associations were chartered by Parliament and specifically directed to support and organize research of broad interest to an entire industry. The earliest research association was chartered in 1917 and supported the scientific instrument industry; the Motor Industry Research Association (MIRA) was formed much later but has been one of the most effective research associations in generating and distributing pertinent information to the British automobile industry. Even though in 1999 all British automobile companies are owned by non-British interests, MIRA continues to be available to companies located in the United Kingdom.

53. "Motor Industry Research Association," 586.

54. Merritt, "Simplification Creates New Problems for Top Management," 238.

55. See *Final Report of the Committee on Industry and Trade*, 214–19.

56. It is interesting to note in this context that the majority of subscribers to the MIRA Bulletin and the RRL's *Technical Reports* were located outside of Britain, principally in France, Germany, and Sweden.

57. Several accounts of the development of antilock braking systems specifically refer to the RRL's work as unimportant and contributing nothing to the development of ABS. Although I think this is untrue, because the RRL played an important role in aiding community formation, it says a lot about the commer-

cial position of the RRL-Jaguar work. From a market standpoint it was simply unsuccessful.

58. Michael Shanks's *The Stagnant Society* touched off a wave of "What's wrong with Britain?" commentaries in the early 1960s. Morgan details this in chapter 6 of *The People's Peace.*

59. "A Retrospective," 7 January 1955, 20.

60. One could scarcely say this was any less true in the United States, where engineers and scientists alike basked in the glow of their World War II victory and the scientific promise of the cold war.

61. "A Retrospective," 12 January 1946, 12.

62. "Institutional Amalgamation," 292.

63. "A Retrospective," 12 January 1946, 12.

64. "Industry and Research," 312.

65. There were several crises of the valuation of the pound and demands by the United States for Britain to adhere to the fixed exchange rates set up by the Bretton Woods agreement. Britain also found herself in the embarrassing position of requiring an International Monetary Fund aid package to support the pound sterling.

66. Lloyd, *Empire, Welfare State, Europe,* 348.

67. Morgan, *The People's Peace,* 235.

68. This change was reflected in the expanded name of the Transport and Road Research Laboratory and its new facility in Crowthorne. The new laboratory fell under the aegis of the Ministry of Transport without DSIR oversight. Meanwhile, the DSIR was reorganized under the Ministry of Technology.

69. "Industrial Research," 416.

4. THE TRACK AND THE LAB

1. Zoeppritz, "An Overview of European Measuring Methods and Techniques."

2. Thomson, "Electrical Units of Measurement," quoted in Wise, introduction, 5.

3. Bradley and Wood, "Some Experiments," 46.

4. Godbey, "The Electric Dynamometer," 1.

5. "Report of Brake Subcommittee No. 3," 329–30.

6. Ibid., 4.

7. Depending on the protocol, both the angular deceleration of the wheel and the linear deceleration of the vehicle could be critical.

8. Sinclair and Gulick, "The Dual Brake Inertia Dynamometer," 253.

9. Temperature was a critical measurement for Johns-Mansville, because this company was one of the largest producers of asbestos products, primarily of relevance to this case as brake linings. A critical feature for brake linings was that their properties be unaffected by high temperature.

10. Sinclair and Gulick, "The Dual Brake Inertia Dynamometer," 253.

11. Ibid., 256.

12. "Report of Brake Sub-Committee No. 3," 17.

13. As a British invention disc brakes became common on production autos in England much earlier than elsewhere. This peculiarity facilitated the work of Lister and Kinchin on the Maxaret system, which was compatible only with disc brakes.

14. "Brakes," 26 November 1958, 495.

15. "Brakes," 14 May 1965, 243.

16. Bielecke and Bethke, "Vergleichsversuche mit Scheiben- und Trommelbremsen," 197–202.

17. For an extensive discussion of the nature of research-technology, see Joerges and Shinn, "A Fresh Look at Instrumentation."

18. Hashimoto, "Theory, Experiment, and Design Practice," 6.

19. See Layton, "Mirror-Image Twins," 562–80. See also Constant, "Scientific Theory and Technological Testability," 183–98.

20. See Milliken and Whitcomb, "Research in Automobile Stability," 287–309.

21. "Ferodo Research Centre," 204.

22. Odier, "A Contribution to the Study of the Dynamics of Vehicles during Braking," 69. See also Odier, "Contribution to the Study of Brakes and the Rational Use of Friction Materials."

23. Odier, Molinier, and Thirion de Briel, "Recent Progress in Braking Tests," 161. See also Odier, "A Dynamometer on Which the Dynamic Behavior of a Passenger Car Can Be Simulated," 7–12; Odier, "Conception et étude d'une nouvelle machine d'essai automobile simulant la tenue sur route," 531–38.

24. Hales, Barter, and Oliver, "Assessment of Vehicle Ride and Handling," 19.

25. Segel, "Theoretical Prediction and Experimental Substantiation of the Response of the Automobile to Steering Control," 310–30.

26. Constant, "Scientific Theory and Technological Testability," 196.

27. Segel and Murphy, "The Variable Braking Vehicle," 276–99.

28. Ibid., 276.

29. Landon, "Model 10 Brake Test Instrument."

30. Hales, Barter, and Oliver, "Assessment of Vehicle Ride and Handling," 15.

31. Spurr, "Subjective Assessment of Brake Performance," 393.

32. Zoeppritz, "An Overview of European Measuring Methods and Techniques."

33. Except perhaps traction control, which is a closely related system developed from ABS technology.

5. FROM THINGS BACK TO IDEAS

1. Frehse, "Fundamentals of Brake Design," 113.

2. Jones, "The Skidding Behaviour of Motor Vehicles," 65–73.

3. See discussion of Jones's paper, "The Skidding Behaviour of Motor Vehicles," especially comments by Ellis, 78.

4. Wind tunnel simulations pose exemplary scale effect problems. The resistance of various shapes to wind does not scale linearly or predictably, which makes testing scale models tricky. Still, for economic and logistical reasons, most airfoils were, in fact, tested as scaled-down models, especially before the era of ubiquitous computer simulations, which were developing simultaneously with ABS. As a result, an array of mathematical approaches which accommodate and extend known experimental values have been generated in aerodynamics research. Aerodynamicists' approaches have been borrowed for many other kinds of scale effects, even when unrelated to aerodynamics.

5. Milliken and Whitcomb, "Research in Automobile Stability," 287.

6. Ibid.

7. In addition to "Research in Automobile Stability," see Milliken and Whitcomb, "Design Implications of a General Theory of Automobile Stability and Control," 367–424.

8. Milliken and Whitcomb, "Research in Automobile Stability," 287.

9. W. F. Milliken, "Author's Replies to Discussion," 409, following Milliken and Whitcomb, "Design Implications."

10. Milliken and Whitcomb, "Design Implications," 367.

11. Milliken, "Author's Replies to Discussion," 419.

12. Milliken and Whitcomb, "Research in Automobile Stability," 298.

13. Ronayne, "Discussion in London," *Institution of Mechanical Engineers, Proceedings of the Automobile Division*, 1956–57, 399, following Milliken and Whitcomb, "Design Implications."

14. It is quite apparent that in the 1950s aeronautical engineering was the sine qua non of mathematical modeling for engineers. Engineers interested in computational mechanics almost always took up aeronautical problems. This was equally true in Europe and the United States. For a case study, see A. Johnson, "From Berkeley to Boeing."

15. Gough, "Discussion in London," *Institution of Mechanical Engineers, Proceedings of the Automobile Division*, 1956–57, 392, following Milliken and Whitcomb, "Design Implications."

16. Milliken, "Author's Replies," 409.

17. Leonard Segel, "Authors' Replies to Discussion," *Institution of Mechanical Engineers, Proceedings of the Automobile Division*, 1956–57, 415, following Segel, "Theoretical Prediction."

18. Ibid. Interestingly, Segel uses the word "instrument" as a verb, writing that the engineers "instrument" the auto, and the automobile is "instrumented."

19. Segel, "Theoretical Prediction," 310.

20. K. N. Chandler, "Theoretical Studies in Braking," 147.

21. Self-energizing brakes use the rotation of the wheel to tighten the brake shoes'

grip on the wheel. Without taking self-energizing into account the calculated torque on the brake would be considerably less than the actual torque.

22. This could be measured on a flywheel or bearing, something where friction with the surface was less of a factor.

23. Ellis, "Communications," 406.

24. Ellis and Sharp, "Measurement of Vehicle Characteristics," 71. It merits note that Robin Sharp was honored in 2008 for his own contributions to vehicle dynamics, partly in the mathematical modeling of motorcycle dynamics.

25. Ellis, "The Dynamics of Vehicles During Braking," 20.

26. Ibid., 21.

27. Stability means the car will continue in the same line in which it is aimed; steerability means that the driver has directional control of the vehicle and can change that line. Clearly these attributes are interrelated, but in the case of antilock systems they are actually different conditions. Rear wheel locking creates instability, while front wheel locking eliminates steerability.

28. Odier, "A Contribution to the Study of the Dynamics of Vehicles during Braking," 70.

29. Odier, "Sur l'influence des caractéristiques générales d'une automobile," 878.

30. Odier, "Road-Holding," 204.

31. With Mitschke as an example, it seems Germans were doing a different kind of work, more closely related to the problems of antilock braking system control. Even more important, Mitschke, the leading vehicle dynamicist, did not publish in English journals, which slowed the diffusion of his work outside Germany. Eventually he became known to the English-speaking world through MIRA translations of his work.

32. Mitschke, "Fahrtrichtungshaltung—Analyse der Theorien," 157–62.

33. Mitschke, "Fahrtrichtungshaltung und Fahrstabilität von vierräderigen Kraftfahrzeugen."

34. Mitschke and Wiegner, "Blockiervorgang eines Gebremsten Rades."

35. His articles in *Automobiltechnische Zeitschrift* show this process. "Fahrtrichtungshaltung—Analyse der Theorien" explains what he thinks are the fatal flaws in earlier models of vehicle dynamics, and "Blockiervorgang eines Gebremsten Rades" shows an array of alternative modeling approaches that he advocates.

36. I discuss Mitschke's connections with German antilock system research in chapter 7.

37. Nader, *Unsafe at Any Speed*, 58–59.

38. Ellis, "Application of Computers," 143.

39. Koiter and Pacejka, "Skidding of Vehicles," 8.

40. Guntur and Ouwerwerk, "Adaptive Brake Control System," 855.

41. Adaptive control and the electronic control of mechanical systems generally were important in many different fields in the 1960s. The concept of adaptive control came to antilock system design from other fields. See Mindell, *Between*

Human and Machine; Mindell, "Automation's Finest Hour"; Mindell, "Engineers, Psychologists, and Administrators."

42. Guntur and Ouwerwerk, "Adaptive Brake Control System," 855.

43. A car that is not slipping at all will have virtually no braking efficiency; this is equivalent to stopping the car by allowing the friction of the engine and road to gradually slow the car to a stop.

44. Guntur and Ouwerwerk, "Adaptive Brake Control System," 858. Also Guntur, "Design Considerations of Adaptive Brake Control Systems," 35.

45. Guntur and Ouwerwerk, "Skid Prediction," 69.

46. Ibid., 69–70.

47. See Koiter and Pacejka, "Skidding of Vehicles," 3. Also Guntur and Ouwerwerk, "Adaptive Brake Control System."

6. LEARNING FROM FAILURE

1. It is not my intention to adjudicate a priority dispute here. Rather, the priority dispute can be used to examine how competing participants in the development of ABS defined what their system did and how that function constituted an antilock system. The acceptance of such definitions is a function of the knowledge community.

2. Interestingly, this question has not faded with time. In 2000 Daimler-Chrysler ran a series of television advertisements in the United States highlighting their role in automotive safety. In these ads, they claim to have "pioneered the first antilock braking systems in 1978." This marketing choice is all the more interesting since Daimler-Benz's new partner at the time, Chrysler, actually introduced an antiskid system to the market seven years before Daimler's. The Chrysler system is not mentioned in the advertisement, nor are the actual producers of the system Daimler is taking credit for: first Teldix, then Bosch.

3. In "How ABS Was Born," 60, Jean-Pierre Gosselin claims that the association with Jensen was disastrous for Dunlop, since the Jensen FF also pioneered four-wheel drive. The marriage of antiskid braking and four-wheel drive was difficult even in the 1980s, and by pioneering two complex innovations on the same vehicle, Dunlop's reputation for reliability was ultimately compromised by outside factors.

4. Note that it has taken nearly twenty years for this level of saturation to be achieved in the United States, where antilock systems are now most common. Only by the late 1990s were antilock systems installed on a majority of new cars in the United States, Europe, and Japan. Daimler claims to have been the first manufacturer to offer ABS as standard equipment, which they did in 1992. Given the number of cars Daimler sells, which is dwarfed by GM's numbers, this claim means nothing in traffic safety terms.

5. Strien, "Trends in the Development," 9.

6. Ibid. When Strien talks about "dosedly-braked," he means modulating the brakes, that is, applying the brakes in doses, as opposed to engaging them fully and constantly. This usage is peculiar to Strien.

7. Nader, *Unsafe at Any Speed*, 58–59.

8. Stewart and Bowler, "Road Testing"; Harned, Johnston, and Scharf, "Measurement of Tire Brake Force Characteristics."

9. Madison and Riordan, "Evolution of the Sure-Track," 895.

10. During World War II, the U.S. Navy contracted Kelsey-Hayes to develop an add-on unit to prevent skidding on carrier-based aircraft. The unit was never produced during the war, but starting in 1949 Kelsey-Hayes and several other companies introduced units that could be retrofit to existing airplane braking systems. See letter from Bureau of Aeronautics to F. H. LeJeune of Hayes Industries, 1944 (no exact date), Bureau of Aeronautics General Correspondence 1943–45, F13–7, Box 936, National Archives, Washington, D.C.

11. Burckhardt and Ostenwall, "Möglichkeiten und Grenzen von Antiblockiersysteme," 13.

12. Madison and Riordan, "Evolution of Sure-Track," 896.

13. "A Review of Anti-Skid Braking," 56.

14. Madison and Riordan, "Evolution of Sure-Track," 898.

15. See chapter 9 for the problem this assumption caused in the long run.

16. "A Review of Anti-Skid Braking," 34.

17. As similar as they are, there was no connection between the names of Ford's and Chrysler's systems.

18. "You Can Steer," 54.

19. Douglas and Schafer, "The Chrysler Sure-Brake," 175.

20. Ibid., 176.

21. Ibid.

22. Ibid., 182.

23. Rae, *The American Automobile Industry*, 138; Wetmore, "Systems of Restraint," 51.

24. Volti, *Cars and Culture*, 118.

25. Wetmore, "Systems of Restraint," 61.

7. *EINES IST SICHER!*

1. "The Little Economic Miracle," 24.

2. This said, electronic fuel injection was a much smaller and less complex device than ABS. Still, the company sold its track record in the adaptive electronic control of a hydraulic system more than the obvious differences between the systems.

3. Kullgatz, "One Thing Is Certain," 32.

4. A. Johnson, "Unpacking Reliability."

5. Bassett, *To the Digital Age;* Braun and Macdonald, *A Revolution in Miniature.*

6. J.-P. Gosselin, "How ABS Was Born," 60.

7. Braun and Macdonald, *A Revolution in Miniature,* 201–2.

8. The mandate reappeared on the 1995 revision of FMVSS-121. It stipulated that in 1997 new tractors were required to have antilock devices; air-braked trailers and single-unit trucks and buses had to comply by 1998; and hydraulic systems had to have antilock devices by 1999. See Insurance Institute for Highway Safety and Highway Loss Data Institute, "Q&A: Antilock Systems for Car, Trucks, Motorcycles" available at www.iihs.org.

9. Originally produced for the U.S. Air Force, in the end these planes were more popular with European NATO members, especially the Federal Republic of Germany. This Lockheed plane, advertised as "a missile with a man in it," had a reputation for being difficult to handle, earning it a whole different category of nicknames, including "the Widowmaker" and "Erdnagel" or "Earth-nail." Nearly a third of 916 German Starfighters crashed. Chuck Yeager also crashed a Starfighter and was seriously injured while trying to set an altitude record in 1963.

10. Kullgatz, "One Thing Is Certain," 31. For the surprisingly complicated merger history of Teldix, see www.rockwellcollins.com.

11. J.-P. Gosselin, "How ABS Was Born," 62.

12. *Elmer A. Sperry Award,* 13.

13. J.-P. Gosselin, "How ABS Was Born," 61.

14. Rixmann, "Neues aus der Industrie," 106. Also see Czinczel and Müller, "Electronic Antiskid System."

15. *Elmer Sperry Award 1993,* 4.

16. Leiber and Limpert, "Ein neuer Weg," 65–70.

17. Leiber and Limpert, "Der Elektronische Bremsregler," 181.

18. Leiber and Limpert, "Ein neuer Weg," 65.

19. Daimler registered ABS as a trademark for the system it was developing with Teldix, or that Teldix was developing using Daimler's resources. When Bosch purchased its interest in Teldix in 1973 it also bought the rights to the trademarked ABS. Today many different systems go by the initials ABS, and Daimler claims that it has never pursued a patent infringement suit on a safety device. Many different manufacturers have licensed various parts of the system Bosch controls, which is also the system on which Daimler holds a trademark.

20. Deceleration was calculated as the change in velocity over the change in time.

21. Leiber and Limpert, "Ein neuer Weg," 67.

22. Mitschke and Wiegner, "Blockiervorgang eines gebremsten Rades," 4.

23. Leiber and Limpert, "Der Elektronische Bremsregler," 183.

24. Rixmann, "Neues aus der Industrie," 106.

25. Leiber and Limpert, "Der Elektronische Bremsregler," 189.

26. For plans to go digital, see ibid., 181. For inadequacies of integrated circuits in the 1960s, see *Elmer Sperry Award 1993,* 10.

27. Leiber and Limpert, "Der Elektronische Bremsregler," 183.

28. Most companies and agencies working on antiskid devices built one of these, which was called a split coefficient track.

29. Czinczel and Müller, "Electronic Antiskid System."

30. Ibid.

31. Leiber, "Der Elektronische Bremsregler," 269.

32. Czinczel and Müller, "Electronic Antiskid System."

33. Leiber, Czinczel, and Anlauf, "Antiblockiersystem für Personenkraftwagen," 72–74.

34. Coefficients of friction are dimensionless scalars.

35. To this day, Teldix, which has been bought and sold several times and is currently a part of Rockwell Collins, is probably best known for the Teldix Space Wheel, a ball-bearing momentum wheel used to stabilize satellites.

36. Mitschke and Wiegner, "Blockiervorgang eines Gebremsten Rades," 4.

37. Ibid., 21.

38. Mitschke and Wiegner, "Simulation von Panikbremsungen," 289.

39. Rixmann, "Neues aus der Industrie," 107.

40. *Elmer Sperry Award 1993*, 11.

41. Burckhardt and Ostenwall, "Möglichkeiten und Grenzen von Antiblockier-system," 13.

42. Engineers at Bosch have a system of prototypes. The first, a machine shop–produced mock-up of a design, is called the A-prototype. As the design is refined and changed the prototype letters advance: B-prototype, C-prototype, and so on. The first mass-produced items are still referred to as prototypes because in some ways they are as experimental and changeable as the first crude model.

43. *Elmer Sperry Award 1993*, 12.

44. Ibid.

45. Leiber and Czinczel, "Four Years of Experience with 4-Wheel Antiskid Brake Systems." While this figure amounts to 50,000 vehicles a year, it is merely a drop in the bucket compared to the number of vehicles produced. Although ABS can be considered the first mass-produced antilock system, it was not until 1994, sixteen years after the introduction of ABS, that a majority of new cars in the United States had antilock brakes. In Europe, Japan, and Australia it has taken until 1999 to achieve this level of market saturation. Ironically, by the time ABS penetration reached 50 percent, Bosch was only the fourth largest producer, having been outpaced by Varity Kelsey-Hayes, ITT (which purchased Teves), and GM's Delphi unit. Blumenstein and Henderson, "GM Skids," B1.

46. ATE *Bremsen-Handbuch*, 3.

47. Ibid.

48. Klein and Fink, "Introduction of Antilock Braking Systems for Cars."

49. Spiegel, *Growth of Economic Thought*, 666.

1. Mahoney, "Finding a History for Software Engineering," 10.
2. Prior to the Stevenson-Wydler Act of 1980 (which was parallel to the Bayh-Dole Act but aimed at research in federal laboratories as opposed to universities), patenting by U.S. federal lab employees was rare and complicated.
3. Giles to McIvor, 12 March 1962, PRO, DSIR 28/406, U.K. Patent Application 31986/63.
4. NRDC Patent Questionnaire, no date, PRO, DSIR 28/406.
5. Ibid.
6. Giles to Cooper, 14 October 1964, PRO, DSIR 28/406.
7. Sefton, Jones, O'dell and Stephens to DSIR, 16 November 1964, PRO, DSIR 28/406.
8. Collins, *Changing Order*, chapter 6.
9. Price, "Is Technology Historically Independent of Science?," 562.
10. Ibid., 561, Price's italics.
11. Garrett, "Engineering Conferences," 497.
12. Daimler claims that it has never challenged a safety-oriented patent.
13. Ferguson, *Engineering and the Mind's Eye*, 23.
14. Ibid., 2–3.
15. Mitcham, *Thinking through Technology*, 224. Mitcham overlooks the less common instances where drawing is used to conceptualize and relate concepts together. Drawing an electrical circuit is a more conceptual exercise than drawing a structure, and circuit design relies on the visual expression of relational concepts in a formal style of drawing.
16. On the other hand, people who build models on paper tend to be much more expensive to employ, so Mitcham's economic argument may be a bit short-sighted.
17. Calvert's *The Mechanical Engineer in America* is largely responsible for outlining the professionalization of engineering as a struggle between school culture and shop culture. It is not my purpose here to argue that this is an inappropriate approach to the development of engineering as a profession, but the relation of mathematics to drawing gets lost when mechanical drawing is considered shop culture and mathematics is placed in the incommensurable category of school culture. In many instances, drawing and mathematics are intertwined for engineers, so that dividing them into two different cultures of engineering makes impossible an examination of the connections between them. One brief example is the importance of shear and moment diagrams in structural analysis. These diagrams are representations of actual physical stresses in a member, as well as being an expression of the differential relationships between the equations of shear and bending moments. The relationship between drawing and

mathematics in these diagrams is the single problem that initially interested me in studying the practice of engineering.

18. An interesting exception is Slaton, *Reinforced Concrete and the Modernization of American Building.*

19. Staudenmeier, *Technology's Storytellers,* 108.

20. In particular, see Layton, "Mirror Image Twins." Also Vincenti, "Control-Volume Analysis."

21. Edwin T. Layton has addressed the issue of engineering as applied science in "Mirror Image Twins." He traces the idea that science begets technology back to the early nineteenth century, although he also points to Vannevar Bush's concepts of pure and applied science as the culmination of this way of thinking. He refers to Bush's oft quoted line, "Basic science is the pacemaker of technological progress" (Bush, *Endless Horizon,* 52), but ultimately dismisses this service-oriented version of engineering as applied science. In "Edison and the Pure Science Ideal" David Hounshell's explanation of Henry Rowland's "Plea for Pure Science" also lends some critical historical sensibility to the relationship of pure and applied science and technology. Paul Forman has also taken on the relationship of science and technology, seeing a major epochal shift occurring since 1980 which reverses the arrow of influence to set technology as a primary shaper of scientific inquiry. See Forman, "The Primacy of Science in Modernity."

22. Pahl and Beitz, *Engineering Design,* 1, italics added.

23. Of course, one could argue that these people, who identify themselves as scientists, are in fact not scientists because the knowledge they produce is not scientific. However, in discussing the development of community, actors' categories are important, so people trained in scientific fields—physics or chemistry in the ABS case—who self-identify as scientists must be considered scientists and their identity contrasted with self-identified engineers. When they refer to theories of vehicle dynamics they claim these theories as engineering theories, or sometimes computational theories. I think Vincenti's claim that engineers create theoretical tools rather than theories is a possible solution to this somewhat semantic issue.

24. Both Vincenti in *What Engineers Know* and Mitcham in *Thinking through Technology* do this in their efforts to lay out epistemologies of engineering. While both admit that the practices of engineers can differ from their schematic representation, they still split theory building from the day-to-day practices of engineers. In examining everyday practices of engineers, one certainly finds many who do not construct theories. However, in the case of ABS the theory builders never formed a distinct group from those developing brake-testing protocols and instruments or those working to develop a commercially viable ABS.

25. Galison, *Image and Logic.*

26. Mitcham, *Thinking through Technology,* 220.

1. Briesch, "Opinion of the European Economic and Social Committee."

2. Aschenbrenner, Biehl, and Wurm, "Schriftenreihe Unfall und Sicherheitsforsch-ung," 65–70; Aschenbrenner, Biehl, Wurm, and Mehr, "Verkehrssicherheit durch bessere Technik?"; Aschenbrenner and Biehl, "Improved Safety through Improved Technological Measures?" An assessment of this study also appears in Wilde, *Target Risk*; however, Wilde is using the study explicitly to make a case for risk compensation, so the assessment is instrumental in his argument.

3. Miller, "Safety Quiz," A1.

4. Organization for Economic Cooperation and Development, Scientific Expert Group, "Behavioural Adaptations to Changes in Road Systems."

5. Priez et al., "How About the Average Driver?"; Grant and Smiley, "Driver Response to Antilock Brakes," 211–20.

6. Highway Loss Data Institute, *Collision and Property Damage.*

7. IIHS press advisory, 16 March 1994.

8. Peltzman, "The Effects of Automobile Safety Regulation," 677–726; Crandall and Graham, "Automobile Safety Regulation."

9. Wilde, "Risk Homeostatis Theory," 89–91; Wilde, "Theory of Risk Homeosta-tis," 209–25. To be clear, Wilde is not claiming that all humans want the same level of risk, but rather that an individual has a preference for a certain level of risk and will modify his or her behavior to maintain that exposure.

10. Insurance Institute for Highway Safety, "Antilock Brakes Don't Reduce Fatal Crashes."

11. Brian O'Neill, quoted in Insurance Institute for Highway Safety, "Antilock Brakes Don't Reduce Fatal Crashes."

12. National Highway Traffic Safety Administration, "Light Vehicle Antilock Brake Systems (ABS) Research."

13. Hertz, "Analysis of the Crash Experience of Vehicles Equipped with All Wheel Antilock Braking Systems (ABS)." See also Farmer, "New Evidence," 361–69. In 2007 the current generation of ABS is often referred to as ABS-8, as this is the current Bosch model, introduced in 2004 on the Saturn Ion. However, as with questions of the "first" ABS, this numbering system is highly subjective. It is also worth mentioning that high-end ABS is now coupled with vehicle traction control and thus constitutes a different artifact.

14. See www.nhtsa.dot.gov.

15. See www.mucda.mb.ca.

16. M. L. Gosselin, Fournier, and Béchard, "Driver Knowledge"; Broughton and Baughan, "The Effectiveness of Antilock Braking Systems in Great Britain," 347–55; "Effectiveness of ABS."

17. Cummings and Grossman, "Antilock Brakes."

Bibliography

- - - - -

ARCHIVES

Public Record Office (PRO), London. Department of Scientific and Industrial
Research (DSIR), Road Research Laboratory (RRL).
United States National Archives. Bureau of Aeronautics (BUAer), General and
Confidential Correspondence (1940–45).

PUBLISHED SOURCES

Aitken, Hugh. *Syntony and Spark: The Origins of Radio.* New York: John Wiley and
Sons, 1976.
Akera, Atsushi. *Calculating a Natural World: Scientists, Engineers and Computers
During the Rise of U.S. Cold War Research.* Cambridge: MIT Press, 2006.
Albert, Daniel. "Order out of Chaos: Automobile Safety, Technology and Society,
1925–1965." Ph.D. diss., University of Michigan, 1997.
"Anti-Skid Device." *Automobile Engineer,* July 1958, 248–54.
Aschenbrenner, K. M., and B. Biehl. "Improved Safety through Improved
Technological Measures? Empirical Studies Regarding Risk Comparison and
Perception in Relation to Antilock Braking." *Challenges to Accident Prevention:
The Issue of Risk Compensation Behavior,* ed. R. M. Trimpop and G. B. Wilde,
81–89. Groningen: Styx, 1994.
Aschenbrenner, K. M., B. Biehl, and B. Wurm. "Schriftenreihe Unfall und
Sicherheitsforschung." *Strassenverkehr* 69 (1988), 65–70.
Aschenbrenner, K. M., B. Biehl, B. Wurm, and G. W. Mehr. "Verkehrssicherheit
durch bessere Technik? Felduntersuchungen zur Risikokompensation am
Beispiel des Antiblockiersystems (ABS)." *Bericht zum Forschungsprojekt 8323*
Bundesanstal fur StraBenwesen Bergisch Gladbach, 1992.
ATE *Bremsen-Handbuch: Berechnung, Funktion, Prüfung, Wartung, and
Umstandsetzung.* Ottobrunn bei München: Autohaus Verlag, 1988.
Baird, Davis. *Thing Knowledge: A Philosophy of Scientific Instruments.* Berkeley:
University of California Press, 2004.
——. "The Thing-y-ness of Things: Materiality and Design. Lessons from
Spectrochemical Instrumentation." *The Empirical Turn in the Philosophy of
Technology,* ed. Anthonie Meijers and Peter Kroes, 99–117. Amsterdam: JAI
Press, 2000.
Baker, James Stannard. "Cases of Skidding Accidents." *Proceedings, First*

International Skid Prevention Conference, 21–26. Washington: Highway Research Board, 1959.

Barwell, F. T. "Friction and Its Measurement." *Metallurgical Reviews* 4, no. 14 (1959), 141–77.

Bassett, Ross. *To the Digital Age: Research Labs, Start-up Companies, and the Rise of MOS Technology*. Baltimore: Johns Hopkins University Press, 2002.

Bell, J. P. "The Motor Vehicle in Relation to Accidents and Injuries." *Journal of Medicine, Science, and the Law* 3, no. 1 (1962), 461.

Bent, Arthur J. "Automatic Brake Control." *Machine Design*, February 1950, 140–42.

——. "No More Skids for Airplanes!" *Society of Automotive Engineers Journal*, January 1950, 63–66.

Bielecke, F.-W., and H.-A. Bethke. "Vergleichsversuche mit Scheiben- und Trommelbremsen." *Automobiltechnische Zeitschrift* 64, no. 7 (1962), 197–202.

Bijker, Wiebe. *Of Bicycles, Bakelite and Bulbs: Towards a Theory of Sociotechnical Change*. Cambridge: MIT Press, 1995.

Blumenstein, Rebecca, and Angelo Henderson. "GM Skids, Slips behind in Antilock-Brake Technology." *Wall Street Journal*, 20 March 1996.

Bradley J., and S. A. Wood. "Some Experiments on the Factors Affecting the Motion of a Four-Wheeled Vehicle When Some of Its Wheels Are Locked." *Proceedings of the Institution of Automobile Engineers* 25 (1930).

"Brakes." *Automobile Engineer*, 26 November 1958.

"Brakes." *Automobile Engineer*, 14 May 1965.

Braun, Ernest, and Stuart Macdonald. *A Revolution in Miniature: The History and Impact of Semiconductor Electronics*. New York: Cambridge University Press, 1982.

Bremsen-Handbuch: Elektronische Brems-Systeme. Ottobrunn: Autohaus Verlag, 1995.

Briesch, Roger. "Opinion of the European Economic and Social Committee on the "Proposal for a Directive of the European Parliament and of the Council Relating to the Protection of Pedestrians and Other Vulnerable Road Users in the Event of a Collision with a Motor Vehicle and Amending Directive 70/156/EEC.' " *Official Journal of the European Union*, 30 September 2003, C234/10–13. http://eur-lex.europa.eu.

Broughton, Jeremy, and Chris Baughan. "The Effectiveness of Antilock Braking Systems in Great Britain." *Accident Analysis and Prevention* 34 (2002), 347–55.

Bucciarelli, Louis L. *Designing Engineers*. Cambridge: MIT Press, 1994.

Buderi, Robert. "In Search of Innovation." *Technology Review* 102, no. 6 (1999), 42–51.

Burckhardt, Manfred. *Fahrwerktechnik: Bremsdynamik und Pkw-Bremsanlagen*. Würzberg: Vogel Buchverlag, 1991.

——. *Fahrwerktechnik: Radschlupf-Regelsysteme*. Würzberg: Vogel Buchverlag, 1993.

Burckhardt, Manfred, and Egon-Christian Glasner von Ostenwall. "Möglichkeiten und Grenzen von Antiblockiersysteme." *Automobiltechnische Zeitschrift* 77, no. 1 (1975), 13–18.

Bush, Vannevar. *Endless Horizon.* Washington: Government Printing Office, 1946.

———. *Science: The Endless Frontier.* Washington: Government Printing Office, 1945.

Calvert, Monte. *The Mechanical Engineer in America.* Baltimore: Johns Hopkins University Press, 1967.

Chandler, Alfred D. *Strategy and Structure: Chapters in the History of Industrial Enterprise.* Cambridge: MIT Press, 1962.

Chandler, K. N. "Theoretical Studies in Braking." *Institution of Mechanical Engineers, Proceedings of the Automobile Division* 4 (1960–61), 147–64.

Chayne, Charles. "Automotive Design Contributions to Highway Safety." *Annals of the American Academy of Political and Social Science* 320 (November 1958), 73–83.

Collier, G. H. "Two Methods of Aircraft Skid Control." Society of Automotive Engineers Paper 580257. New York: Society of Automotive Engineers, 1958.

Collins, Harry. *Changing Order: Replication and Induction in Scientific Practice.* London: Sage, 1985.

Constant, Edward W. *Origins of the Turbojet Revolution.* Baltimore: Johns Hopkins University Press, 1980.

———. "Scientific Theory and Technological Testability: Science, Dynamometers, and Water Turbines in the 19th Century." *Technology and Culture* 24 (1983), 183–98.

Crandall, Robert, and John Graham. "Automobile Safety Regulation and Offsetting Behavior." *American Economic Review* 74, no. 2 (1984), 328–31.

Cummings, Peter, and David Grossman. "Antilock Brakes and the Risk of Driver Injury in a Crash: A Case-Control Study." *Accident Analysis and Prevention* 39 (2007), 995–1000.

Czinczel, Armin, and P. Müller. "Electronic Antiskid System: Performance and Application." Society of Automotive Engineers Paper 725046. New York: Society of Automotive Engineers, 1972.

Davis, Michael. *Thinking Like an Engineer.* Oxford: Oxford University Press, 1998.

Davis, Stacy C. *Transportation Energy Data Book,* edition 20. Washington: U.S. Department of Energy, 2000. Available at www.cta.ornl.gov.

Dodgson, Mark. *Technological Collaboration in Industry.* London: Routledge, 1993.

Dosi, Giovanni. "Technological Paradigms and Technological Trajectories." *Research Policy* 11 (1982), 147–62.

Douglas, J. W., and T. C. Schafer. "The Chrysler Sure-Brake—the First Production Four Wheel Antiskid System." Society of Automotive Engineers Paper 710248. New York: Society of Automotive Engineers, 1971.

Downey, Gary L., Arthur Donovan, and Timothy J. Elliott. "The Invisible Engineer: How Engineering Ceased to Be a Problem in Science and Technology Studies." *Knowledge and Society* 8 (1989), 189–216.

"Effectiveness of ABS and Vehicle Stability Control Systems." *Research Report 04/01 of the Royal Automobile Club of Victoria Ltd.* Victoria, Australia: RACV, 2004.

Ellis, J. R. "Application of Computers." *Automobile Engineer*, April 1970, 143–44.

——. "Communications." *Institution of Mechanical Engineers, Proceedings of the Automobile Division*, 1956–57, 406–7.

——. "The Dynamics of Vehicles During Braking." *Institution of Mechanical Engineers, Symposium on the Control of Vehicles*, 20–29. London: Institution of Mechanical Engineers, 1963.

Ellis, J. R., and R. S. Sharp, "Measurement of Vehicle Characteristics for Ride and Handling." *Institution of Mechanical Engineers: Instrumentation and Test Techniques for Motor Vehicles* 182, Part B, 71–81. London: Institution of Mechanical Engineers, 1967.

Elmer A. Sperry Award 1993 for Advancing the Art of Transportation. Program. Warrendale, Pa.: Society of Automotive Engineers, 1994.

Farmer, Charles. "New Evidence Concerning Fatal Crashes of Passenger Cars before and after Adding Antilock Braking Systems." *Accident Analysis and Prevention* 33 (2001), 361–69.

Ferguson, Eugene. *Engineering and the Mind's Eye.* Cambridge: MIT Press, 1992.

"Ferodo Research Centre." *Automobile Engineer*, June 1959.

Fiala, E. "Die Wechselwirkung zwischen Fahrzeug und Fahrer." *Automobiltechnische Zeitschrift* 69 (1967), 222–79.

Final Report of the Committee on Industry and Trade. London: HMSO, 1929.

Forbes, T. W. "Driver Knowledge, Judgement and Responses in Causation and Control of Skidding." *Proceedings, First International Skid Prevention Conference*, 51–58. Charlottesville: Virginia Council of Highway Investigation and Research, 1959.

Forman, Paul. "The Primacy of Science in Modernity, of Technology in Postmodernity, and of Ideology in the History of Technology." *History and Technology* 23, nos. 1–2 (2007), 1–152.

Frehse, A. W. "Fundamentals of Brake Design." Society of Automotive Engineers Paper 300016. New York: Society of Automotive Engineers, 1930.

Galison, Peter. *Image and Logic.* Chicago: University of Chicago Press, 1997.

Garrett, T. K. "Engineering Conferences." *Automobile Engineer*, December 1961.

Garrott, Riley W., and Elizabeth Mazzae. "An Overview of the National Highway Traffic Safety Administration's Light Vehicle Antilock Brake Systems Research Program." Society of Automotive Engineers Paper 1999–01–1286. Warrendale, Pa.: Society of Automotive Engineers, 1999.

Giles, C. G., and G. Grime. "The Skid-Resisting Properties of Roads and Tyres." *Institution of Mechanical Engineers, Proceedings of the Automobile Division*, 1954–55, 19–30.

Giles, C. G., and Barbara E. Sabey. "Skidding as a Factor in Accidents on the

Roads of Great Britain." *Proceedings, First International Skid Prevention Conference*, 27–42. Charlottesville: Virginia Council of Highway Investigation and Research, 1959.

Glanville, W. H. "Summary." *Proceedings, First International Skid Prevention Conference*, ix–xi. Charlottesville: Virginia Council of Highway Investigation and Research, 1959.

Godbey, Roy S. "The Electric Dynamometer." Society of Automotive Engineers Paper 480152. New York: Society of Automotive Engineers, 1948.

Goodhart, A. L. "Statistics and Road Accidents." *Journal of Medicine, Science and the Law* 3 (1962), 432–39.

Gosselin, Jean-Pierre. "How ABS Was Born." *Automobile Year* 34 (1985–86), 57–64.

Gosselin, Martin Lee, Pierre-Sébastien Fournier, and Isabelle Béchard. "Driver Knowledge and Beliefs about Antilock Braking Systems." *Transportation Research Record 1779*, Paper no. 01–3293. Université Laval, Québec, 2001.

Grant, B., and A. Smiley. "Driver Response to Antilock Brakes: A Demonstration of Behavioural Adaptation." *Proceedings of the Canadian Multidisciplinary Road Safety Conference VIII*, 14–16 June 1993, 211–20.

Gunsaulus, A. C. "Rolling Wheels Gather No Skids." *Society of Automotive Engineers Transactions* 61 (1953), 189–95.

Guntur, Ramachandra Rao. "Design Considerations of Adaptive Brake Control Systems." Society of Automotive Engineers Paper 741082. New York: Society of Automotive Engineers, 1974.

Guntur, Ramachandra Rao, and H. Ouwerkerk. "Adaptive Brake Control System." *Institution of Mechanical Engineers, Automobile Division Proceedings* 186/68 (1972), 855–80.

———. "Laboratory Testing of Antiskid Devices." *Journal of Automotive Engineering*, February 1973, 22–25.

———. "Skid Prediction." *Vehicle Systems Dynamics* 1 (1972), 67–88.

Hales, F. D., N. F. Barter, and R. J. Oliver. "Assessment of Vehicle Ride and Handling." *Instrumentation and Test Techniques for Motor Vehicles* 182 Part B, 14–22. London: Institution of Mechanical Engineers, 1967.

Halsey, M. "Design Cars for Poor Drivers." *Society of Automotive Engineers Journal*, June 1948, 17–20.

Harned, J. L., L. E. Johnston, and G. Scharf. "Measurement of Tire Brake Force Characteristics as Related to Wheel Slip (Antilock) Control System Design." Society of Automotive Engineers Paper 690214. New York: Society of Automotive Engineers, 1969.

Hashimoto, Takehiko. "Theory, Experiment, and Design Practice: The Formation of Aeronautical Research, 1909–30." Ph.D. diss., Johns Hopkins University, 1991.

Hertz, Ellen. "Analysis of the Crash Experience of Vehicles Equipped with All Wheel Antilock Braking System (ABS)—A Second Update Including Vehicles with Optional ABS." *DOT Report 809 144*. Washington: National Highway Transportation Safety Administration, 2000.

Highway Loss Data Institute. *Collision and Property Damage Liability Losses of Passenger Cars with and without Antilock Brakes*. Report A-41. Arlington, Va.: Highway Loss Data Institute, 1994.

Hill, D. W. *Co-operative Research in Industry*. London: Hutchinson's Scientific and Technical Publications, 1947.

Hobsbawm, Eric. "Introduction: Inventing Traditions." *The Invention of Tradition*, ed. Eric Hobsbawm and Terence Ranger, 1–14. New York: Cambridge University Press, 1992.

———. "Mass Producing Traditions: Europe, 1870–1914." *The Invention of Tradition*, ed. Eric Hobsbawm and Terence Ranger, 263–308. New York: Cambridge University Press, 1992.

Hounshell, David. "Edison and the Pure Science Ideal in 19th-Century America." *Science* 207, no. 4431 (1980), 612–17.

Hughes, Thomas H. *Networks of Power: Electrification in Western Society, 1880–1930*. Baltimore: Johns Hopkins University Press, 1983.

"Industrial Research." *The Engineer*, 23 March 1960.

"Industry and Research." *The Engineer*, 5 April 1946.

"Institutional Amalgamation." *The Engineer*, 29 March 1946.

Insurance Institute for Highway Safety. "Antilock Brakes Don't Reduce Fatal Crashes. People in Cars with Antilocks at Greater Risk—But Unclear Why." News release. 10 December 1996. Available at www.iihs.org.

Joerges, Bernward, and Terry Shinn, "A Fresh Look at Instrumentation: An Introduction." *Instrumentation between Science, State and Industry*, ed. Bernward Joerges and Terry Shinn, 1–13. Dordrecht: Kluwer, 2001.

Johnson, Ann. "From Berkeley to Boeing: Civil Engineers, the Cold War and the Development of Finite Element Analysis." *Growing Explanations: Historical Perspectives on the Sciences of Complexity*, ed. M. Norton Wise, 133–58. Durham, N.C.: Duke University Press, 2004.

———. "From Dynamometers to Simulations: Transforming Brake Testing Technology into Antilock Braking Systems." *Instrumentation between Science, State and Industry*, ed. Bernward Joerges and Terry Shinn, 199–218. Dordrecht: Kluwer, 2001.

———. "Unpacking Reliability: Robert Bosch GmbH and the Construction of ABS as a Reliable Product." *History and Technology* 17 (2001), 249–70.

Johnson, P. S. *Co-operative Research in Industry*. New York: Wiley, 1973.

Jones, G. "The Skidding Behaviour of Motor Vehicles." *Institution of Mechanical Engineers, Proceedings of the Automobile Division*, 1962–63, 65–73.

Keyser, J. H. "Electrical Prerotation of Landing Gear Wheels." *Electrical Engineer* 67 (December 1959), 454–59.

Kidder, Tracy. *Soul of a New Machine*. New York: Avon Books, 1982.

Kinchin, J. W. "Disc Brake Development and Antiskid Braking Devices." Society of Automotive Engineers Paper 620525. New York: Society of Automotive Engineers, 1962.

Klein, Hans Christof, and Werner Fink. "Introduction of Antilock Braking Systems for Cars." Society of Automotive Engineers Paper 741084. New York: Society of Automotive Engineers, 1974.

Kline, Ronald. "Construing 'Technology' as 'Applied Science': Public Rhetoric of Scientists and Engineers in the United States, 1880–1925." *Isis* 86, no. 2 (1995), 194–221.

Knight, Frank. *Risk, Uncertainty and Profit.* 1921; reprint, Chicago: University of Chicago Press, 1971.

Koiter, W. T., and H. B. Pacejka. "Skidding of Vehicles Due to Locked Wheels." *Institution of Mechanical Engineers, Automobile Division Proceedings* 183/3H (1968–69), 3–17.

Kuhn, Thomas. "Second Thoughts on Paradigms." *The Structure of Scientific Theories,* ed. Frederick Suppe. Urbana: University of Illinois Press, 1974.

——. *The Structure of Scientific Revolutions.* 2nd ed. Chicago: University of Chicago Press, 1970.

Kullberg, Gösta, Olle Nordström, and Göran Palmkvist. "Antilock Braking System for Passenger Cars: Development of a Brake System Giving Yaw Stability and Steerability during Emergency Braking." *Statens-väg och transportforsknings institut,* Report no. 100A.

Kullgatz, Dietrich. "One Thing Is Certain: Bosch ABS Series Production." *Bosch History Magazine,* 2003.

Kummer, H. W., and W. E. Meyer. "Verbesserter Kraftschluß zwischen Reifen und Fahrbahn—Ergebnisse einer neuen Reibungstheorie." Part 1. *Automobiltechnische Zeitschrift* 69, no. 8 (1967), 245–51.

Kummer, H. W., and W. E. Meyer. "Verbesserter Kraftschluß zwischen Reifen und Fahrbahn—Ergebnisse einer neuen Reibungstheorie." Part 2. *Automobiltechnische Zeitschrift* 69, no. 11 (1967), 382–86.

Landon, G. W. "Model 10 Brake Test Instrument." Society of Automotive Engineers Paper 700374. New York: Society of Automotive Engineers, 1970.

Layton, Edwin T., Jr. "Mirror-Image Twins: The Communities of Science and Technology in Nineteenth Century America." *Technology and Culture* 12 (1971), 562–80.

Leiber, Heinz. "Der Elektronische Bremsregler und seine Problematik." *Automobiltechnische Zeitschrift* 74, no. 7 (1972), 269–77.

Leiber, Heinz, and Armin Czinczel. "Four Years of Experience with 4-Wheel Antiskid Brake Systems." Society of Automotive Engineers Paper 830481. New York: Society of Automotive Engineers, 1983.

Leiber, Heinz, Armin Czinczel, and Jürgen Anlauf. "Antiblockiersystem für Personenkraftwagen." *Bosch Technische Berichte* 7 (1980), 65–94.

Leiber, Heinz, and Wolfgang D. Limpert. "Der Elektronische Bremsregler." *Automobiltechnische Zeitschrift* 71, no. 6 (1969), 181–89.

———. "Ein neuer Weg zur Verhütung des Blockierens von Kraftfahrzeugrädern durch adaptive Bremsschlupfregelung." *Automobil-Industrie*, April 1968, 65–70.

Lister, R. D. "Brake Performance Measurement." *Automobile Engineer* 49, no. 7 (1959).

Lister, R. D., and R. N. Kemp. "Skid Prevention." *Automobile Engineer*, October 1958, 382–91.

Lister, R. D., and H. J. H. Starks. "Experimental Investigations on the Braking Performance of Motor Vehicles." *Institution of Mechanical Engineers, Proceedings of the Automobile Division*, 1954–55, 31–44.

Lister, R. D., and R. G. Stevenson. "Fifth Wheel for Measuring Speed and Braking Distance." *Motor Industries Research Association Bulletin* 4, 1952.

"The Little Economic Miracle." *Bosch History Magazine*, 2001, 24–25.

Lloyd, T. O. *Empire, Welfare State, Europe: English History 1906–92*. Oxford: Oxford University Press, 1993.

Madison, R. H., and Hugh E. Riordan. "Evolution of the Sure-Track Brake System." Society of Automotive Engineers Paper 690213. New York: Society of Automotive Engineers, 1969.

Mahoney, Michael S. "Finding a History for Software Engineering." *Annals of the History of Computing* 26, no. 1 (2004), 8–19.

May, G., ed. *Encyclopedia of American Business History and Biography: The Automobile Industry 1920–1980*. New York: Facts on File, 1989.

Meijers, Anthonie, and Peter Kroes, eds. *The Empirical Turn in the Philosophy of Technology*. Amsterdam: JAI Press, 2000.

Merritt, H. E. "Simplification Creates New Problems for Top Management." *The Engineer*, 24 February 1950.

Miller, Krystal. "Safety Quiz: Insurance Claims Data Don't Show Advantage of Some Auto Devices." *Wall Street Journal*, 17 March 1994.

Milliken, William F., and David A. Whitcomb. "Design Implications of a General Theory of Automobile Stability and Control." *Institution of Mechanical Engineers, Proceedings of the Automobile Division*, 1956–57, 367–424.

———. "Research in Automobile Stability and Control in Tyre Performance: A General Introduction to a Programme of Dynamic Research." *Institution of Mechanical Engineers, Proceedings of the Automobile Division*, 1956–57, 287–309.

Mills, J. P., Jr., and W. B. Shelton. "Virginia Accident Information Relating to Skidding." *Proceedings, First International Skid Prevention Conference*, 9–20. Charlottesville: Virginia Council of Highway Investigation and Research, 1959.

Mindell, David A. "Automation's Finest Hour: Bell Labs and Automatic Control in

World War II." *Institute of Electrical and Electronic Engineers Control Systems* 15 (December 1995), 72–80.

———. *Between Human and Machine: Feedback, Control and Computing before Cybernetics.* Baltimore: Johns Hopkins University Press, 2002.

———. "Engineers, Psychologists, and Administrators: Control Systems Research in Wartime." *Institute of Electrical and Electronic Engineers Control Systems* 15 (August 1995), 91–99.

Mitcham, Carl. *Thinking through Technology: The Path between Engineering and Philosophy.* Chicago: University of Chicago Press, 1994.

Mitschke, Manfred. "Fahrtrichtungshaltung—Analyse der Theorien." *Automobiltechnische Zeitschrift* 70, no. 5 (1968), 157–62.

———. "Fahrtrichtungshaltung und Fahrstabilität von vierrädigen Kraftfahrzeugen." *Deutsche Kraftfahrtforschung und Straßenverkehrstechnik,* book 135, 1960.

Mitschke, Manfred, and Peter Wiegner. "Blockiervorgang eines gebremsten Rades." *Proceedings, 1970 Fédération Internationale des Sociétés d'Ingénieurs des Techniques de l'Automobile, World Automotive Congress* 17.3.C, 1970, 5–25.

———. "Simulation von Panikbremsungen mit verschiedenen Blockierverhinderern auf Fahrbahnen geteilter Griffigkeit." Part 1. *Automobiltechnische Zeitschrift* 77, no. 10 (1975), 289–93.

———. "Simulation von Panikbremsungen mit verschiedenen Blockierverhinderern auf Fahrbahnen geteilter Griffigkeit." Part 2. *Automobiltechnische Zeitschrift* 77, no. 11 (1975), 328–32.

Mom, Werner. "An Improved Safety Device for Preventing Jamming of the Wheels of Automobiles when Stopped." U.K. Patent 382,241, 20 October 1932.

Morgan, Kenneth O. *The People's Peace: British History 1945–1989.* Oxford: Oxford University Press, 1990.

Morison, Elting. *From Know-How to Nowhere.* New York: Signet Press, 1977.

"Motor Industry Research Association." *The Engineer,* 28 June 1946.

Mowery, David C., and Nathan Rosenberg. *Technology and the Pursuit of Economic Growth.* New York: Cambridge University Press, 1989.

Moynihan, Daniel Patrick. "Epidemic on the Highways." *The Reporter* 20, no. 9 (1962).

Nader, Ralph. *Unsafe at Any Speed.* New York: Grossman, 1965.

National Highway Traffic Safety Administration. "Light Vehicle Antilock Brake Systems (ABS) Research." Washington: National Highway Traffic Safety Administration, 2003. www.nrd.nhtsa.dot.gov.

Noble, David F. *America by Design.* New York: Knopf, 1977.

Nordström, Olle. "Provning av antilåssystem: Vinterprov 1985 för värdering av förslag till revision av ECE Reg 13, Annex 13." Väg-och Trafik-Institutet Report no. 304. 1986.

Normann, O. K. "Braking Distances of Vehicles from High Speeds and Tests of Friction Coefficients." *Public Roads* 27, no. 8 (1953), 159–69.

———. "Today's Speeds." *Proceedings, First International Skid Prevention Conference,* 5–20. Charlottesville: Virginia Council of Highway Investigation and Research, 1959.

Odier, Jean. "Conception et étude d'une nouvelle machine d'essai automobile simulant la tenue sur route." *Journal de la Société des Ingénieurs de l'Automobile,* October 1968, 531–38.

———. "Contribution to the Study of Brakes and the Rational Use of Friction Materials." *Société Française des Mécaniciens Bulletin No. 16.* Paris: Société Française des Mécaniciens, 1955.

———. "A Contribution to the Study of the Dynamics of Vehicles during Braking." *Proceedings, Fédération Internationale des Sociétés d'Ingénieurs des Techniques de l'Automobile, World Automotive Congress IX,* 69–88. London: Institution of Mechanical Engineers, 1962.

———. "A Dynamometer on Which the Dynamic Behavior of a Passenger Car Can Be Simulated." *Proceedings of the Institution of Mechanical Engineers,* 1972.

———. "Road-Holding: Braking and Traction." *International Automobile Safety Conference Compendium,* 204–9. New York: Society of Automotive Engineers, 1970.

———. "Sur l'influence des caractéristiques générales d'une automobile et du facteur vitesse sur la stabilité en freinage." *Comptes-rendus Academie des Sciences* 249 (1959), 878–80.

Odier, Jean, P. Molinier, and J. Thirion de Briel. "Recent Progress in Braking Tests by Use of a Car Dynamometer." *1976 Institution of Mechanical Engineers Conference on Road Vehicles,* 161–68. London: Institution of Mechanical Engineers, 1976.

Oetzel, J. G. "When Are Brakes Adequate?" *Society of Automotive Engineers Transactions* 63 (1955), 243–52.

Ooudshorn, Nelly, and Trevor Pinch, eds. *How Users Matter.* Cambridge: MIT Press, 2003.

Oppenheimer, Paul. "Safer Braking Systems." *Institution of Mechanical Engineers, Proceedings of the Automobile Division* 187 (October 1973), 87–97.

Organization for Economic Cooperation and Development, Scientific Expert Group. "Behavioural Adaptations to Changes in Road Systems." Paris: OECD Road Transport Research, 1990.

Pahl, Gerhard, and Wolfgang Beitz. *Engineering Design: A Systematic Approach.* New York: Springer-Verlag, 1988.

Palmer, F. "Friction." *Scientific American,* February 1951.

Peltzman, Sam. "The Effects of Automobile Safety Regulation." *Journal of Political Economy* 83, no. 4 (1980), 677–726.

Petroski, Henry. *Design Paradigms: Case Studies of Error and Judgment in Engineering.* New York: Cambridge University Press, 1994.

——. *To Engineer Is Human: The Role of Failure in Successful Design.* New York: Vintage Books, 1992.

Phillips, A. *Technology and Market Structure.* Lexington, Mass.: Heath Lexington Books, 1971.

Pickering, Andrew, ed. *Science as Practice and Culture.* Chicago: University of Chicago Press, 1992.

Polanyi, Michael. *Personal Knowledge.* Chicago: University of Chicago Press, 1962.

Price, Derek J. de Solla. "Is Technology Historically Independent of Science? A Study in Statistical Historiography." *Technology and Culture* 6 (1965), 553–68.

——. "The Parallel Structures of Science and Technology." *Science in Context,* ed. Barry Barnes and David Edge, 164–76. Cambridge: MIT Press, 1982.

Priez, Alain, et al. "How About the Average Driver in a Critical Situation: Can He Really Be Helped by Primary Safety Improvements?" *Proceedings of the 13th International Technical Conference on Experimental Safety Vehicles,* Document DOT HS 807 990. Washington: U.S. Department of Transportation, 1991.

Pyatt, Edward. *The National Physical Laboratory: A History.* Bristol, England: Adam Hilger, 1983.

Radder, Hans, ed. *The Philosophy of Scientific Experimentation.* Pittsburgh: University of Pittsburgh Press, 2003.

Rae, John B. *The American Automobile Industry.* Boston: Twayne, 1984.

Rammert, Werner. "Two Styles of Knowing and Knowledge Regimes: Between 'Explication' and 'Exploration' under Conditions of 'Functional Specialization' or 'Fragmental Distribution.' " Berlin Technical University Technology Studies Working Papers TUTS-WP-3–2004. Berlin: Technische Universität Berlin, 2004. www2.tu-berlin.de.

Randle, J. N. "From Sidecars to Silicon—A History of Innovation at Jaguar Cars." Institution of Mechanical Engineers, Proceedings of Section D, Transport Engineering 201 no. D3 (1987), 155–66.

"Report of Brake Subcommittee No. 3: Construction and Operation of Brake Dynamometers." Society of Automotive Engineers Paper 530130. New York: Society of Automotive Engineers, 1953.

"Report of Subcommittee B: Accidents and the Human Element in Skidding." *Proceedings, First International Skid Prevention Conference,* xxvi–xxix. Charlottesville: Virginia Council of Highway Investigation and Research, 1959.

Research on Road Safety. London: HMSO, 1963.

"A Retrospective." *The Engineer,* 12 January 1946.

"A Retrospective." *The Engineer,* 7 January 1955.

"A Review of Anti-Skid Braking." *Automotive Engineering,* July 1975.

Rixmann, Werner. "Neues aus der Industrie: Das Bremsregler-System ABS von DB-Teldix." *Automobiltechnische Zeitschrift* 73, no. 3 (1971), 106–7.

Road Research Laboratory Technical Report 122. Slough: Road Research Laboratory, 1952.

Robert Bosch GmbH. *Bremsanlagen für Kraftfahrzeuge*. Stuttgart: Robert Bosch GmbH, 1994.

Rosenberg, Nathan. *Exploring the Black Box*. New York: Cambridge University Press, 1994.

———. *Inside the Black Box*. New York: Cambridge University Press, 1981.

———. *Perspectives on Technology*. New York: Cambridge University Press, 1976.

Rosenberg, Nathan, and Walter Vincenti. *Britannia Bridge: The Generation and Diffusion of Knowledge*. Cambridge: MIT Press, 1978.

Schon, Donald. "Fear of Innovation." *Science in Context*, ed. Barry Barnes and David Edge, 290–302. Cambridge: MIT Press, 1982.

———. *The Reflective Practitioner*. New York: Basic Books, 1983.

Segel, Leonard. "Theoretical Prediction and Experimental Substantiation of the Response of the Automobile to Steering Control." *Institution of Mechanical Engineers, Proceedings of the Automobile Division*, 1956–57, 310–30.

Segel, Leonard, and Ray W. Murphy. "The Variable Braking Vehicle: Concept and Design." *Proceedings of the First International Conference on Vehicle Mechanics*, 276–99. Amsterdam: Swets and Zeitlinger, 1968.

Self, Thomas L. "Breaking Skids Betters Braking." *Aviation Week*, 4 June 1951, 49–53.

Shanks, Michael. *The Stagnant Society*. Harmondsworth, England: Penguin Books, 1963.

Shelburne, Tilton E. Introduction. *Proceedings, First International Skid Prevention Conference*, vii–ix. Charlottesville: Virginia Council of Highway Investigation and Research, 1959.

Sinclair, David, and W. F. Gulick. "The Dual Brake Inertia Dynamometer—A New Tool for Brake Testing." Society of Automotive Engineers Paper 630464. New York: Society of Automotive Engineers, 1963.

Slaton, Amy. *Reinforced Concrete and the Modernization of American Building, 1900–1930*. Baltimore: Johns Hopkins University Press, 2001.

Spiegel, Herman. *Growth of Economic Thought*. Durham, N.C.: Duke University Press, 1991.

Spurr, R. T. "Subjective Assessment of Brake Performance." *Automobile Engineer*, September 1965, 393–95.

Starks, H. J. H. "Experimental Investigation of the Braking Performance of Motor Vehicles." *Institution of Mechanical Engineers, Proceedings of the Automobile Division*, 1954–55, 31–44.

Staudenmeier, John. *Technology's Storytellers: Reweaving the Human Fabric*. Cambridge: MIT Press, 1985.

Stewart, Edwin E., and Lauren L. Bowler. "Road Testing of Wheel Slip Control Systems in the Laboratory." Society of Automotive Engineers Paper 690215. New York: Society of Automotive Engineers, 1969.

"Stopping without Skidding." *Engineering*, 12 April 1963.

Strien, Hans. "Trends in the Development of Vehicle Brakes and Anti-Skid Braking

Devices in Europe." Society of Automotive Engineers Paper 610171. New York: Society of Automotive Engineers, 1961.

Vincenti, Walter. "Control-Volume Analysis: A Difference in Thinking between Engineering and Physics." *Technology and Culture* 23 (April 1982), 145–74.

———. *What Engineers Know and How They Know It.* Baltimore: Johns Hopkins University Press, 1990.

Volti, Rudi. *Cars and Culture.* Baltimore: Johns Hopkins University Press, 2004.

Wehner, Bruno. "Accidents Involving Slippery Road Conditions in Germany." *Proceedings, First International Skid Prevention Conference*, 45–50. Charlottesville: Virginia Council of Highway Investigation and Research, 1959.

Wengenroth, Ulrich. *Enterprise and Technology: The German and British Steel Industries, 1865–1895.* New York: Cambridge University Press, 1994.

Wetmore, Jameson. "Redefining Risks and Redistributing Responsibilities: Building Networks to Increase Automobile Safety." *Science, Technology and Human Values* 29, no. 3 (2004), 377–405.

———. "Systems of Restraint: Redistributing Responsibilities for Automobile Safety in the United States Since the 1960s." Ph.D. diss., Cornell University, 2003.

"Wheel Lock Inhibited Braking." *The Engineer*, 20 June 1958.

Whyte, R. R. *Engineering Progress through Trouble.* London: Institution of Mechanical Engineers, 1975.

Wilde, Gerald J. S. "Risk Homeostasis Theory: An Overview." *Injury Prevention* 4 (1998), 89–91.

———. *Target Risk.* Toronto: PDE Publications, 1994.

———. "Theory of Risk Homeostatis: Implications for Safety and Health." *Risk Analysis* 2 (1982), 209–25.

Wise, M. Norton. Introduction. *The Values of Precision*, ed. M. Norton Wise, 3–13. Princeton: Princeton University Press, 1995.

"You Can Steer with 4 Wheel Skid Control." *Automotive Engineering*, October 1970, 54.

Zoeppritz, Hanns. "An Overview of European Measuring Methods and Techniques." *Transportation Research Record* no. 621. Washington: National Academy of Sciences, 1977.

Index

Chrysler, xii, 104, 107–8, 114–16, 135; Chrysler Imperial, 111, 113. *See also* Sure-Brake
Citroën, 31
closure, 10, 16
Collins, Harry, 19, 148
community of practitioners, 19
computer, 88, 90, 91, 92, 97; in automobiles, 76–77, 99; simulations, xi, 64, 88, 97, 130
Constant, Edward, 34–35, 73, 76
Consumers Union, xiii
Cooper, A. F., 55, 143
cost, 30, 31, 70–71, 81, 107–8, 109, 115, 126, 133, 157
Cornell Aeronautical Laboratory, 73–74, 75, 86–88, 90, 92–93, 95, 98, 109
crash avoidance, 26
Czinczel, Armin, 123

Daimler-Benz, xiii, 16, 30, 33, 121, 123, 127, 131, 146–47
Data General, 17
DBA. *See* Bendix
deaths, traffic, 115
deceleration, 126–27
Department of Scientific and Industrial Research (U.K.), 39, 44, 54, 57, 60, 142
design, ix, 3–4, 6, 9, 15, 16, 122, 135; analysis of, 10–12; knowledge, 7, 15, 136, 150–51, 154; mathematical models in, 88, 151; process of, 10, 11, 43, 152; as social activity, 8, 17, 34, 57–58
Digital Equipment Corporation, 17
digital logic circuit, 126, 132
directional control, xii, 27, 126, 130, 146
directional stability, xii, 27, 92, 96, 111, 112, 113, 126
disc brakes, 29, 30, 49, 50, 69–70, 105, 108, 133
"disciplinary matrix," 34

Doppler radar, 55
Dosi, Giovanni, 35
Douglas, J. W., 112–13
drawing, as communication, 150–51
driver education, 28, 39, 46, 48, 52; for ABS, 162–64
drivers, 77–79, 83, 112, 115, 158; behavior of, 96, 102; error of, 42, 45, 162; psychology of, x, 42, 43, 45, 78; response to ABS, 158–62; skill of, 110, 120; test, 72, 77–79, 87, 114
drum brakes, 27, 30, 49, 50, 70
Dunlop, xi, 29, 30, 48, 50–51, 58, 71, 79, 108, 119, 145
dynamometer, xi, 65–67

efficiency, of brakes, 49, 122–23, 125
electronic control, xii, 20, 108, 110, 112, 113, 121, 123, 131, 133, 145, 146
electronics, 108–9, 112, 118, 119, 120, 121, 122, 130, 133, 153–54
Ellis, J. R., 92–94
Elmer Sperry Award, 18, 20, 104, 137, 148
empirical data, xi, 73, 77–79, 87–88, 92
emissions standards, 117, 129
engineering: design practice in, 6–12, 17; professional status of, 138, 149; publications, xiii, 21, 31, 33, 53, 56, 86, 94–95, 97–98, 105, 115, 124–25, 137–39, 143, 145–49, 165; studies of, 6, 154
engineering management, in Britain, 57–58
ENIAC-11, 93
"epidemic on the highways," 25
epistemology, xvii, xviii, 4, 6–7, 153
Europe, 3, 33
experiments, 65–66, 92
expertise, 19, 20, 28, 31, 35, 121, 148

failure: of ABS, 81, 110, 130, 131, 145, 163; of mathematical models, 92

Federal Motor Vehicle Safety Standard, 121, 120

Ferguson, Eugene, 9, 13, 150

Ferodo, 74, 78, 94

Fiat, 31

fifth wheel apparatus, 46, 68, 70, 79, 113

Fink, Werner, 134

FISITA (International Congress of Automotive Engineers), 31, 129

Ford, xii, 104, 107–11, 112, 113; Scorpio, 133; Thunderbird, 111

Forkenbrock, Gary, 163

FORTRAN, 98

France, 31

Frehse, A. H., 85

friction, 26, 27, 38, 41, 43–44, 53–54, 67, 71, 74, 91, 110, 124, 126, 127–29

fuel injection, 118, 121, 129, 131

Galison, Peter, 153

General Motors, 57, 77, 107

Geneva, 69, 133

Germany, 25. See also West Germany

Gerstenmeier, Hans Jürgen, 18, 119–33, 146, 148

Giles, Cyril George, 44, 52–56, 61, 63, 141–43

Glanville, W. H., 28, 37, 42, 43, 46, 63

Goodyear, 27, 48, 55, 142

Gosselin, Jean-Pierre, 123, 137

government funding of research, 29, 39, 56–61

Great Britain, x, 25, 27, 28, 39–41, 86, 88, 94–95; industrial decline in, 56–61

Gulick, W. F., 66–67

Guntur, Ramachandra Rao, 99–101, 129, 136

gyroscopes, 30, 121, 125, 145

handling of vehicles, xi, 64, 71, 72–73, 74, 75, 86, 97

Harmondsworth, 39–40

Hayes Industries, 27. See also Kelsey-Hayes

Heidelberg, 30, 120, 123, 146

Highway Loss Data Institute (U.S.), 159

Highway Research Board (U.S.), 38

Highway Safety Research Institute (U.S.), 75–77, 89–90

Howe, H. A., 54

"how" questions, in ABS design, 6, 7, 8–9, 64

Hydro-Aire, 27, 48, 109

hydroplaning, 41

industry collectives, 57

Institut für Fahrzeugtechnik, 95

Institution of Mechanical Engineers (U.K.), 33, 45, 59, 86, 89–90, 95

instruments for measurement. See metrology

Insurance Institute for Highway Safety (U.S.), xiv, 159–61

integrated circuit, 126, 132

intellectual property law, 142

interaction problem, xi, 28, 29, 39–40, 42–43, 58, 60, 63, 140

International Congress of Automotive Engineers (FISITA), 31, 129

International Skid Prevention Conference: in 1958, x, 19, 28, 37, 82; in 1977, 80

Jaguar, 30, 50–51, 58, 79, 105, 145

Japan, 3, 56

Jensen FF, 30, 58, 105, 145

Johns-Manville, 66–67

Jones, G., 85

Jonner, Wolf-Dieter, 18, 119–33, 146–48

Kelsey-Hayes, xii, 107, 108–11, 112, 113, 119, 162

safety (*continued*)
135; of ABS, 157–62; mandates for, 115, 163
safety features, 114–15
scale, effects of, 40, 71–72, 86
Schafer, T. C., 112–13
Schon, Donald, 14–15, 135
SCS (Lucas), 33
seat belts, 115, 157, 160
Sefton, Jones, O'dell, and Stephens, 54, 56, 142
Segel, Leonard, 75–76, 79–80, 89–90
self-energizing brakes, 91
semiconductors, 119–20, 121, 135
sensors, 32, 48, 111, 112, 113, 127, 128
Sharp, Robin, 92
Shelburne, Tilton E., 37
Sheridan, Thomas B., 165
Shinn, Terry, 16–17, 72, 141
sideslip, 126
Sinclair, David, 66–67
skidding: defining problem of, 2, 26, 41–42, 67–68; statistics on, 25, 29, 30, 41
Slough, England, 37, 54
Snow, C. P., 60–61
Society of Automotive Engineers (SAE), 65, 104, 106, 116, 122, 163
sociology, historical, 4
speed, highway, 42
Sperry Award, 18, 20, 104, 137, 148
Spurr, R. T., 78–79
stability. *See* directional stability
standards: for emissions, 117; performance, 21, 64–65
Starfighter (F104 jet), 30, 120, 144
"Starrion," 1, 164
Statens-väg och trafikinstitut (VTI), 32
statistics on skidding, 25, 29, 30, 41
Staudenmaier, John, 152
steering, xii, 89

stopping distance. *See* braking distance
Strien, Hans, 31, 106–7
Stuttgart, 18, 121, 123, 137
Sure-Brake, 111–16, 117, 125, 130, 134
Sure-Track Brake System, 108–11, 117, 134
suspension, 91
Sweden, 32, 41, 81, 130

taxi studies (in Munich), 158–60, 165
Technische Hochschule, Braunschweig, 95, 129
Technische Hogeschool, Delft, 98–101, 129, 136
technological fix, xiv, 3, 26, 42
technology, studies of, 151–52
Teldix, GmbH, 18, 20, 30–31, 33, 82, 104, 118–31, 135, 144–48
Teledyne, 109
Telefunken, 120
test drivers, 72, 77–79, 87, 114
testing: facilities for, 32, 40, 64–65, 68, 127; methods of, 46–47, 57, 65–67
Teves, 31, 33, 104, 106, 118, 133–36, 162
Texas Instruments, 109
theories, 21, 145; scientific, 7, 34, 151–52; of vehicle dynamics, 79, 87, 89, 124
thing knowledge, 7, 13, 21, 154
Thomson, William (Lord Kelvin), 64
tires: blowouts of, 27; design of, x, 28, 37, 42, 44–45, 52, 77
trading zones, 153
traditions of practice, 34
traffic: deaths, 115; increase in, 41
transistors, 121
Transport Canada, 164
Two Cultures debate, 60

United States, xii, 3, 25, 28, 56
University of Michigan, 77. *See also* Highway Safety Research Institute
University of Virginia, x, 19, 28, 37

Ann Johnson is an associate professor of
history and philosophy at the University
of South Carolina.

Library of Congress Cataloging-in-Publication Data
Johnson, Ann
Hitting the brakes : engineering design and
the production of knowledge / Ann Johnson.
p. cm.
Includes bibliographical references and index.
ISBN 978-0-8223-4526-8 (cloth : alk. paper)
ISBN 978-0-8223-4541-1 (pbk. : alk. paper)
1. Automobiles—Antilock brake systems.
2. Automobiles—Design and construction.
3. Automobiles—Technological innovations.
4. Technological innovations—Social aspects.
5. Knowledge management. I. Title.
TL269.3.J64 2009
629.2'4609—dc22 2009030094